TECHNICAL
REPORT

Methodology for Improving the Planning, Execution, and Assessment of Intelligence, Surveillance, and Reconnaissance Operations

Sherrill Lingel, Carl Rhodes, Amado Cordova,
Jeff Hagen, Joel Kvitky, Lance Menthe

Prepared for the United States Air Force

Approved for public release; distribution unlimited

PROJECT AIR FORCE

The research reported here was sponsored by the United States Air Force under Contracts F49642-01-C-0003 and FA7014-06-C-0001. Further information may be obtained from the Strategic Planning Division, Directorate of Plans, Hq USAF.

Library of Congress Cataloging-in-Publication Data

Methodology for improving the planning, execution, and assessment of intelligence, surveillance, and reconnaissance operations / Sherrill Lingel ... [et al.].
p. cm.
Includes bibliographical references.
ISBN 978-0-8330-4171-5 (pbk. : alk. paper)
1. United States. Air Force—Intelligence Service. 2. Military intelligence—United States. I. Lingel, Sherrill Lee. II. Rand Corporation.

UG633.M345 2007
358.4'134320973—dc22

2007045929

The RAND Corporation is a nonprofit research organization providing objective analysis and effective solutions that address the challenges facing the public and private sectors around the world. RAND's publications do not necessarily reflect the opinions of its research clients and sponsors.

RAND® is a registered trademark.

© Copyright 2008 RAND Corporation

All rights reserved. No part of this book may be reproduced in any form by any electronic or mechanical means (including photocopying, recording, or information storage and retrieval) without permission in writing from RAND.

Published 2008 by the RAND Corporation
1776 Main Street, P.O. Box 2138, Santa Monica, CA 90407-2138
1200 South Hayes Street, Arlington, VA 22202-5050
4570 Fifth Avenue, Suite 600, Pittsburgh, PA 15213-2665
RAND URL: http://www.rand.org/
To order RAND documents or to obtain additional information, contact
Distribution Services: Telephone: (310) 451-7002;
Fax: (310) 451-6915; Email: order@rand.org

Preface

The U.S. Air Force has faced challenges in recent conflicts dealing with emerging, fleeting targets that expose themselves to detection and attack for short periods. As these targets may be vulnerable for only a few minutes, response must be quick. A key part of an effective response to these stressing targets is an intelligence, surveillance, and reconnaissance (ISR) system that has appropriate sensors at the correct location when targets are exposed. To enable this, the ISR planning process must appropriately prioritize many competing tasks and, at the same time, allow flexible, real-time changes to the plan with a minimum of delay and friction.

Lessons learned from operations in Afghanistan and Iraq have indicated that commanders are often unaware of how their ISR assets are being employed and that they are perhaps not being used to their full potential. An end-to-end assessment process is needed that can improve daily ISR planning and platform employment as well as assure the commander that the objectives are being effectively supported by ISR assets under his or her control. To enable this, an ISR assessment process must appropriately integrate a large quantity of both quantitative data and often-piecemeal qualitative judgments from sources with vastly differing perspectives. Furthermore, this process must not be manpower intensive.

This report presents work conducted for a fiscal year 2005 study, "Tasking and Employing Intelligence, Surveillance, and Reconnaissance Assets to Support Effects-Based Operations," on three methodologies that collectively improve the planning, tasking, and employment of ISR assets. It also includes an assessment of how well each strategy performed. We propose improvements in the ISR assessment process through better utilization of existing strategies-to-tasks frameworks, standardized mandatory feedback formats, and better use of limited ISR Division (ISRD) resources. In addition, we discuss the utility of automated systems to reduce the ISR assessment workload and the need for joint and Air Force doctrinal reform to enable effective ISR assessment. Although this report is concerned mainly with Air Force platforms and the Air Operations Center (AOC), most of the concepts discussed here apply equally well at the joint forces commander (JFC) level and to intelligence-gathering platforms operated by other services or at the national level.

We also present ideas for improving ISR collection planning and execution through implementation of a strategies-to-tasks framework for collection planning. In addition, we explore the benefits of function-based collection prioritization.

In Chapters Four and Five of this report, a methodology is described. The methodology uses a quantitative, analytical approach to assess the costs and benefits of proposed ISR collection strategies. The analytic framework is divided into two efforts. The first focuses on the deliberate planning process, while the second looks at the employment or execution of a collection strategy. The report is written for an audience that has subject-matter understanding comparable to that of an intelligence officer.

The research reported here was sponsored by the Commander, Pacific Air Forces; the Director of Intelligence, Headquarters, Air Combat Command; and the Director of Intelligence, Surveillance, and Reconnaissance, Deputy Chief of Staff for Air and Space Operations. The work was conducted within the Aerospace Force Development Program of RAND Project AIR FORCE.

RAND Project AIR FORCE

RAND Project AIR FORCE (PAF), a division of the RAND Corporation, is the U.S. Air Force's federally funded research and development center for studies and analyses. PAF provides the Air Force with independent analyses of policy alternatives affecting the development, employment, combat readiness, and support of current and future aerospace forces. Research is conducted in four programs: Aerospace Force Development; Manpower, Personnel, and Training; Resource Management; and Strategy and Doctrine.

Additional information about PAF is available on our Web site: http://www.rand.org/paf/

Contents

Figures

Tables

Summary

This report presents work conducted for a fiscal year 2005 Project AIR FORCE study, "Tasking and Employing Intelligence, Surveillance, and Reconnaissance Assets to Support Effects-Based Operations." Three methodologies are presented that collectively improve the planning, tasking, and employment of ISR assets. The report also includes an assessment of how each strategy performed.

We examine existing joint and Air Force doctrine along with Air Force tactics, techniques, and procedures to understand the role of assessment procedures in the overall intelligence process. ISR assessment techniques, employed by various Air Force units, are examined to determine how these assessments are implemented during current intelligence operations. This report suggests a number of ideas to help improve the ISR assessment process, including utilization of strategies-to-tasks frameworks, standardizing and mandating feedback, and automating certain processes to better utilize ISR Division resources (p. 39). The focus of this work is on Air Force processes and procedures, but other components, joint forces commands, regional combatant commands, and national intelligence organizations could apply many of these same concepts.

Identifying and analyzing the system and operational trade-offs necessary to ensure effective allocation of limited ISR resources against different target sets are complex and difficult tasks. Intelligence officers do not currently have a means to evaluate the costs and benefits of a particular ISR collection strategy. In this report, we describe (1) a new strategies-to-tasks framework for using ISR assets; (2) organizational, training, and doctrinal modifications to improve ISR assessments; and (3) new models to improve future ISR utilization analyses. We also suggest how these new concepts for command and control of ISR forces can be integrated into training for ISR specialists as well as for potential Joint Forces Air Component Command (JFACC) commanders.

In this report, we describe a model developed to quantify the effectiveness of alternative ISR collection strategies to satisfy the range of ISR requirements found in a major theater conflict. The model's analytic framework is divided into two sections: the first focuses on the planning process of building collection "decks," and the second focuses on assessing the execution of the plans in a simulated environment using different collection strategies. Note that we use the term "deck" here to describe the planned collection schedule—e.g., target 1 will be collected at time 12 by asset 4.

Our modeling efforts are applicable to a range of conflict scenarios, but because of time constraints, we focused on a single scenario.[1] Current ISR forces are used as a baseline for this analysis. The flexibility of this modeling framework is demonstrated by examining the effects of employing different sensors and platforms in the same scenario. Results are classified and, therefore, not included in this document.

To assist in moving ISR planning and execution forward from a fixed target and deliberate planning focus to one centered on emerging targets, we propose enhancing the collection-management process with a strategies-to-tasks and utility-based framework. By linking the top-level commander's guidance, operational objectives, and tasks to specific collections and by employing relative utilities, planning for the daily intelligence collections and real-time retasking for ad hoc ISR targets will be enhanced. When current tools are modified to provide this information, planners will be able to link individual collections to top-level operational objectives for better decisionmaking and employment optimization of collection assets. Similarly, in the AOC, intelligence officers will be better able to deal with time-sensitive, emerging targets by rapidly comparing the value of an ad hoc collection with the value of the collection opportunities already planned. To efficiently respond to the ISR demands posed by the rapidly changing battlefields of the future, this more capable decisionmaking framework will ensure the best use of our limited intelligence assets.

The availability of timely and accurate intelligence is critical in both peacetime and wartime. To ensure efficient use of our limited intelligence assets, an end-to-end assessment process is needed to monitor and evaluate daily operations. To date, the majority of ISR assessments have focused on using statistics from the tactical level (e.g., sorties flown and percentage of planned images collected). The question of whether the ISR system is satisfying the commander's intent has gone largely unanswered by these statistical assessment methods.

A summary of our primary observations and recommendations on ISR assessment includes the following:

- An ISR assessment process is critical for determining how well ISR is supporting campaign objectives (pp. 28–29).
- Poor performance by the ISR system can affect the conduct of the entire campaign (p. 9).
- Air Force and joint doctrine provide little or no guidance on how to perform ISR assessment, only directing that it should be done. Air Force Operational Tactics, Techniques, and Procedures (AFOTTP) 2-3.2 provides, by far, the most detailed and useful guidance on ISR assessment. This guidance along with recent work by the Air Combat Command and current combatant command best practices should be utilized in a bottom-up manner to form the next revision of Air Force ISR doctrine. Joint discussions should also be held to compare techniques across services in preparation for joint doctrine revisions (pp. 13, 17–19, 35).
- Adopting a strategies-to-tasks framework for collection planning at the Joint Task Force (JTF) level will enable much more useful strategic and operational ISR assessments because ISR tasks will be clearly connected to campaign objectives and accompanied by measures of effectiveness (pp. 32–37).

[1] The analytic approach is best suited for characterizing major combat operations rather than subsequent stability operations.

- Standardized, joint manuals for the delineation of measurable ISR tasks should be written. Essential elements of information and observables should be generated and disseminated by the Air Combat Command/A2 or Joint Functional Component Command ISR using best practices from current efforts in this area by the various combatant commands and components (p. 35).
- As applicable, quantitative ISR performance data should be collected and processed using database management systems (p. 37).
- JTF J-2 staff and/or the ISR Division in the AOC should develop and disseminate standard Web-based assessment forms for all requestors and users of ISR-generated intelligence (p. 38).
- JTF and component commanders should mandate feedback on ISR performance from all requestors and users of ISR-generated intelligence. Service and joint doctrine as well as training curricula should reflect this requirement (pp. 37, 38).
- Prior to operations, senior members of the JTF and JFACC intelligence staff should plan to elicit feedback from their respective commanders on ISR's contribution toward achieving objectives (p. 38).

Acknowledgments

Col. Mark Kipphut, Pacific Air Forces, was the sponsor of this work at the project's start. He provided numerous insights that helped initiate this area of research and helped to refine our ideas on the topic. His enthusiasm and knowledge contributed immensely to this work. Subsequently, Col. Martin Neubauer took command of the Pacific Air Forces A-2 branch. He has also provided useful insights on the tasking and assessment of intelligence, surveillance, and reconnaissance.

Lt. Col. Timothy McCaig, Lt. Col. Patrick Flood, and Ken Ayers, Pacific Air Forces, served as action officers for this research. They worked tirelessly to arrange numerous visits for our project team, to provide insights into our results, and to gather data for our modeling efforts.

Brig. Gen. Thomas Wright and Gary Harvey, Air Combat Command, hosted a data-gathering visit by our research team and helped arrange visits for this research. Thanks to their help, we were able to participate in a session writing the concept of operations for assessing ISR operations. Maj. Valerie Champagne of the Air Combat Command Intelligence Squadron also served as a useful source of information on ISR assessment. Brig. Gen. James Poss, Air Combat Command, also provided useful insights about this work regarding the proposed utility framework.

At RAND, we would like to thank Myron Hura and Bruce Bennett for reviewing the document and Terri Perkins, Lisa Lewis, and Jane Siegel for all of their hard work in editing and formatting this document. Jack Gibson provided suggestions that improved the substance and style of this document.

We also thank Capt. Peter Halsey at the 613 Air Operations Center/IRDO for reviewing the document.

Abbreviations

ACC	Air Combat Command
ACF	analysis, correlation, and fusion
AFDD	Air Force Doctrine Document
AFOTTP	Air Force Operational Tactics, Techniques, and Procedures
AIS	Air Intelligence Squadron
AOC	Air Operations Center
AOD	Air Operations Directive
ATO	air tasking order
BDA	Battle or Bomb Damage Assessment
CA	Combat Assessment
CENTAF	Central Command Air Forces
CENTCOM	Central Command
COD	Combat Operations Division
COM	collection operations model
COMINT	communications intelligence
CONEX	concepts of execution
CONOPS	concept of operations
CRT	collection requirements tool
ECEF	earth centered, earth fixed
EEI	essential element of information
ELINT	electronics intelligence
ENU	east north up
EO	electro-optical

ERP	effective radiated power
EW	early warning
FLOT	forward line of troops
GIQE	general image-quality equation
GMTI	ground moving target indicator
GSD	Ground Sample Distance
HRR	high-range resolution
IA	image analyst
IMINT	imagery intelligence
IPB	intelligence preparation of the battlespace
IR	infrared
ISAR	inverse SAR
ISARC	Intelligence, Surveillance, and Reconnaissance Cell
ISR	intelligence, surveillance, and reconnaissance
ISRD	ISR Division
JAOC	Joint Air Operations Center
JAOP	Joint Air Operations Plan
JCMB	Joint Collection Management Board
JFACC	Joint Forces Air Component Command
JFC	joint forces commander
JIPCL	Joint Integrated Prioritized Collection List
JIPTL	Joint Integrated Prioritized Target List
JTF	Joint Task Force
LNO	liaison officer
LOS	line of sight
MOE	measure of effectiveness
MTF	Modulation Transfer Function
MTI	moving target indicator
NIIRS	National Imagery Interpretability Rating Scale
OAT	Operational Assessment Team

OEF	Operation Enduring Freedom
OIF	Operation Iraqi Freedom
PACAF	Pacific Air Forces
PACOM	Pacific Command
PED	processing, exploitation, and dissemination
PIR	priority intelligence requirement
PRISM	Planning Tool for Resource Integration, Synchronization, and Management
ROC	receiver operating characteristic
RSTA	reconnaissance, surveillance, and target acquisition
SAM	surface-to-air missile
SAR	synthetic aperture radar
SEAS	System Effectiveness Analysis Simulation
SIDO	senior intelligence duty officer
SIGINT	signals intelligence
SNR	signal-to-noise ratio
TCPED	tasking, collection, processing, exploitation, and dissemination
TTP	tactics, techniques, and procedures
USAF	U.S. Air Force
USAFE	U.S. Air Forces in Europe
WMD	weapons of mass destruction

CHAPTER ONE

Introduction

Throughout the 1990s, the U.S. Air Force (USAF) greatly increased the number of operational surveillance and reconnaissance sensors and its ability to process data from these sensors, e.g., Distributed Common Ground Stations, in support of operations across a wide range of conflicts. However, along with the increased number of sensors comes an increase in the complexity of the command and control operations needed to prosecute either planned for or emergent battlefield targets. This problem has been compounded by an increased use of mobile systems by our adversaries in recent conflicts, especially high-value assets such as air defenses and surface-to-surface missiles.

To improve performance against emerging targets, the process of planning and executing strike operations has employed dedicated aircraft in "standby" orbit(s) until targets appear. Allocating forces in this manner can be easily accommodated using the traditional air tasking order (ATO) planning process and allows strike assets to respond quickly to time-critical taskings. Such a strategy has little downside when an excess of strike capabilities exists. Many air power experts claim that such a situation existed for USAF strike assets during recent conflicts, including Operations Allied Force and Enduring Freedom.

In contrast, intelligence, surveillance, and reconnaissance (ISR) assets are widely seen as high-demand/low-density (HD/LD) assets. This view is true for all recent conflicts and holds for coverage of both fixed and emerging targets. Often, retasking a sensor to cover a new emerging situation necessitated that another, potentially important, tasking could not be satisfied, resulting in an opportunity cost to the commanders. USAF commanders from recent conflicts claim that current ISR asset apportionment tools are ineffective in making rational trades between conflicting demands for ISR coverage.[1]

In this report, we present alternative methods to (1) approach the tasking and command and control process and (2) assess the outcome of different information collection strategies. As part of our research, we developed new assessment techniques and operational strategies to improve the command and control process for assigning ISR assets in dynamic environments. We also suggest tools to assist commanders of ISR assets in their decisions regarding allocating and retasking ISR assets. We focused on traditional target sets against adversaries whose behavior is well understood. This approach would be problematic for scenarios in which we lack knowledge of the adversary's behavior and strategy, and it would result in a low probability of successful collection. However, any approach may prove difficult under these circumstances. Also, in some situations, current ISR capabilities are insufficient to deliver the desired infor-

[1] It appears that for nontraditional target sets, such as terrorists, the process may also be improved by collection managers and analysts gaining subject-matter expertise on the target set.

mation. In this report, we focus on cases in which intelligence assets at hand may be able to provide the necessary data.

Current modeling tools at RAND or elsewhere are not well suited to carry out this analysis; so new tools were developed based on the logic incorporated in past ISR analyses at RAND. This work also involved understanding organizational issues; for this background we relied upon interviews with USAF personnel who participated in recent conflicts.

This report begins with a review of current joint and Air Force doctrines and procedures in planning, tasking, and assessing ISR operations. In Chapter Three, we briefly discuss some lessons learned in employing ISR assets in recent conflicts and recommend ways to improve the assessment of ISR utilization. These improvements entail employment of a strategies-to-tasks utility-based framework for tasking ISR assets. In Chapter Four, we explain the analytic framework we developed to examine the costs and benefits of alternative collection strategies, for which we developed two models. The first is described in Chapter Four and simulates the deliberate planning process. The second, discussed in Chapter Five, simulates the employment of ISR assets in theater. Chapter Six looks at assessing ISR operations produced by the model. Finally, Appendix A presents National Imagery Interpretability Rating Scale tables for imagery analysts, and Appendix B discusses the effects of geometry on employing ISR sensors.

CHAPTER TWO

Doctrine and Procedures

This chapter begins by discussing the current process of generating ISR collection "decks" and compares this process with the targeting process used by the strategy and plans divisions of the Air Operations Center (AOC). Note that we use the term "deck" to describe the planned collection schedule—e.g., target 1 will be collected at time 12 by asset 4. Shortfalls are identified in the current process. Next, doctrinal guidance at the joint and Air Force levels on the intelligence cycle is described. Particular attention is focused on the assessment steps in the cycle that allow examination and subsequent improvement of ISR employment. This chapter assumes a fair degree of reader knowledge on the organization of ISR among different commands and at headquarters.

Generating Collection Decks

The joint forces commander (JFC)[1] is charged with allocating his or her ISR resources and requests intelligence community resources to efficiently achieve the campaign objectives. Current joint doctrine for allocating wartime ISR resources starts with the commander's critical information requirements to support an overall strategy. Those requirements considered the most important are the priority intelligence requirements (PIRs). The pieces of information critical to addressing the PIRs are called essential elements of information (EEIs), and it may be necessary to gather a number of EEIs to answer all aspects of a given PIR. Each EEI may have specific observables tied to satisfying its requirement.

Each component, including the air component, is going through the process of generating a Component Integrated Prioritized Collection List. In addition, in the AOC, the Combat Plans Division is generating collection requirements to support ongoing strike operations, which are presented in the Joint Integrated Prioritized Target List (JIPTL). A simplified, non-doctrinal view of these processes is shown in Figure 2.1. In the AOC, the output of the process forms the basis of the Air Component's inputs for the joint collection-management process. The integrating step of collection management for all forces is performed at the Joint Task Force (JTF) level or may be delegated to a particular service component (Joint Pub 2-01).

The JTF collection manager is tasked with converting the intelligence requirements into collection requirements to form the Joint Integrated Prioritized Collection List (JIPCL), select-

[1] According to Joint Pub 2-01, *joint forces commander* is a general term applied to a combatant commander, subunified commander, or joint task force commander authorized to exercise combatant command or operational control over a joint force.

Figure 2.1
Integrating Collection Requirements from Multiple, Disparate Sources

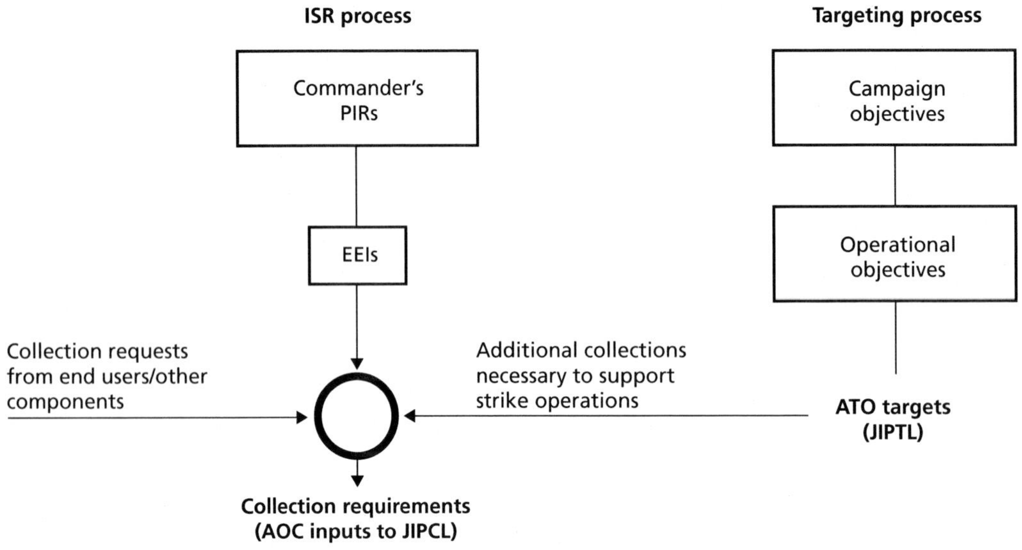

RAND *TR459-2.1*

ing specific collections that will service the EEIs and thus fulfill the PIRs.[2] At the joint level, the collection-management process involves integrating and prioritizing requirements from all components. Guided by an intelligence strategy, the collection manager must make the best use of limited ISR assets while trading off requirements from various sources and satisfying the challenging time constraints associated with wartime operations. The collection manager can allocate forces that are organic to the JTF but is also able to make requests for services from national agencies. Joint doctrine specifies that only those collection requirements that cannot be satisfied by organic assets should be forwarded for potential collection by other systems. An example of this situation might be collections beyond the reach of airborne ISR assets that could be filled by national technical means.

To accomplish this matching of requirements, collections, and assets, joint doctrine states that the decision regarding which requirements should be satisfied with the limited assets is reached via prioritization. That prioritization is assigned "based on the commander's guidance and the current situation" (Joint Pub 2-01, p. III-12). In most cases, a Joint Collection Management Board (JCMB) will be convened by the collection manager and serve as a mechanism for combining and prioritizing the intelligence needs of the various components and the JFC. The JCMB can either be located in a JTF or at the relevant Unified Combatant Command.

Once the requirements are prioritized, subject-matter experts determine collections that answer the EEIs for priority requirements. The ultimate output of this process is the JIPCL. ISR assets are then tasked to satisfy the JIPCL by collecting data on the targets that will satisfy as many requirements as possible during the planning process, with emphasis on those with the highest priority. Typically, a certain percentage of possible collections are allocated to priority number 1, a lower fraction to number 2, and so on until all the possible collections have

[2] While the JIPCL is not approved terminology, we use it here to describe a comprehensive, prioritized collection list of all collection targets for that day's tasking.

been planned or the entire JIPCL has been collected. A certain number of collection "slots" are also left open on each platform to allow for potential ad hoc requests or unplanned collection opportunities.

During the planning phase of ISR operations, the collection manager has a difficult job. He or she begins with the highest-priority requirements and determines how the existing assets can satisfy those requirements. Collection system effectiveness is determined by analyzing the capability and availability of existing assets to collect against a specific set of requirements. The proper asset for collection against a given requirement is weighed against the range to target, timeliness, weather, and geography. Those requirements given a low priority may simply fall off the collection list. For example, Battle or Bomb Damage Assessment (BDA) was assigned a low priority during the combat operations associated with Iraqi Freedom. As a result, very few BDA collections were achieved.[3] After the campaign, some senior USAF leaders claimed that the lack of BDA was a shortfall of the ISR system, rather than recognizing that the situation was the result of a low priority assigned to BDA collections and a lack of ISR assets given the large number of requirements.

The requirement prioritization process described in Joint Pub 2-01 not only addresses the importance of a collection; there is also consideration of target dynamics in the prioritization process. For example, p. III-12 states, "Collection requirements that are not time-sensitive may initially be submitted at lower priorities in the expectation that such requirements may be satisfied during complementary collection operations." This statement implies that time-sensitive collections are assigned a higher priority than would otherwise be the case simply so they are accomplished in a timely fashion. This "gaming" of the priority system is not the most transparent method of accounting for target dynamics.

In practice, operators attempt to use their best judgment in prioritizing new time-critical targets with respect to existing collections. However, at times, the guidance provided in the reconnaissance, surveillance, and target acquisition (RSTA) annex of the ATO is not sufficiently detailed to allow informed decisions to be made by operators at disparate locations.

Shortfalls in the Current Process

In the process just described, collections appearing on the JIPCL are ultimately derived from the JFC's intent. However, once on the JIPCL, it is difficult to trace any individual collection back to the effect that is to be achieved with the collection. The Planning Tool for Resource Integration, Synchronization, and Management (PRISM), the collection-management software currently employed at the Pacific Command (PACOM), allows operators to associate collections with PIRs. However, according to users of the system, a detailed understanding of the role of the collection in satisfying the PIR is not included in PRISM.

Few, if any, written linkages exist between top-level priorities and individual collections. In addition, the reasoning process behind collection decisions is often spread through multiple staff organizations in multiple components. As a result, it becomes difficult to identify ties between the top-level strategies and the collection tasks that help to enact those strategies for ISR operations. Furthermore, with the relative importance of requirements distinguished only by their position in the prioritized ranking, there is insufficient information to make informed

[3] Personal communications with Air Combat Command personnel, May 2003.

trade-offs between collections. Such shortfalls cause difficulties in both the deliberate planning and dynamic retasking processes.

In a paper published by the Air War College, then–Lt. Col. Daniel Johnson (2004) recognized this problem and proposed implementing a strategies-to-tasks framework (see Thaler, 1993) for linking the JFC's guidance to specific tasks via operational objectives. The strategies-to-tasks process starts with broad, campaign-level objectives and links them to operational activities and finally to tactical tasks. By using this framework, it should be easier to understand the contribution of individual collections with respect to the JFC's guidance and to help guide the prioritization process. This process, in turn, should help speed the retasking of ISR assets, because the trade-offs between targets are more readily apparent (Johnson, 2004). Making intelligent decisions about retasking collection assets is currently difficult because it is hard to unravel what is lost at the strategic level by not satisfying a particular collection requirement. However, there are other shortfalls in the current process when it comes time to execute a day's planned operations.

Such a strategies-to-tasks planning mechanism is already firmly entrenched in the Strategy Division of the AOC (AFOTTP 2-3.2). The Air Operations Directive (AOD) provides guidance for those in the Combat Plans cell in a strategies-to-tasks framework. At times, guidance for ISR tasks to be accomplished are also placed in the AOD, but there is no standardized mechanism for incorporating this information in the existing computational tools used by collection managers. While there are personnel from the ISR Division (ISRD) of the AOC assigned to the Strategy Division considering these issues, better automation could help these divisions work together more efficiently.

Ideally, a commander should be able ensure that his or her PIRs are being satisfied with the appropriate level of effort rather than simply prioritizing individual collections. It should be transparent throughout the chain of command why certain collections are being performed and others are not. It should be possible to determine when to replace planned collections with ad hoc collections. Such a method should allow for separating the importance of any given collection requirement from the likelihood of successfully collecting against that requirement. The utility of a successful collection and the probability of a successful collection are two distinct and separable terms. For these reasons, we intend to expand upon the strategies-to-tasks framework laid out by Johnson (2004) to help senior leadership and ISR operators to better plan and execute ISR operations under a framework of centralized control and decentralized execution.

It should be noted that this proposed framework is simply a tool to help operators and decisionmakers with the planning and execution of ISR operations. Like any good tool, this framework is not intended to replace good military judgment. We envision situations in which this framework may not be consulted or employed, for a variety of reasons. However, it could be a useful addition to current processes and procedures for planning and executing ISR operations. Development of the strategies-to-tasks framework is described in Chapter Three.

Assessing the Effects of ISR Asset Employment

We now turn our attention to the topic of assessing ISR operations. Joint Pubs 2-0 and 2-01 are the main contributors in the joint arena, while a variety of Air Force concepts of operations (CONOPSs); tactics, techniques, and procedures (TTP); instructions; and unit-level procedures have been developed to fill most of the remaining doctrinal gaps.[4] This section briefly highlights most of the major sources of guidance, from the broadest to the most specific.

Doctrinal Guidance on the Intelligence Cycle

As with most processes supporting military operations, gathering information with an ISR system[5] is a cycle of setting objectives, prioritizing to accommodate limited capacity, executing plans, and disseminating and analyzing the results. According to Air Force doctrine, the cycle includes nine specific steps, beginning with the formation of a commander's guidance and concluding with the employment of ISR assets in the operational mission (AFOTTP 3-3.6). Figure 2.2, taken from AFOTTP 3-3.6 (previously AFDD 2-5.2), illustrates the various steps in this cycle.

A critical step of this process is the evaluation of the effectiveness of ISR. This step is represented in the figure by the "Evaluate" box and the arrow labeled "Feedback" connecting the consumers of the information back to the planners. Unfortunately, unlike combat assessment,

Figure 2.2
Air Force Doctrinal Description of ISR Process

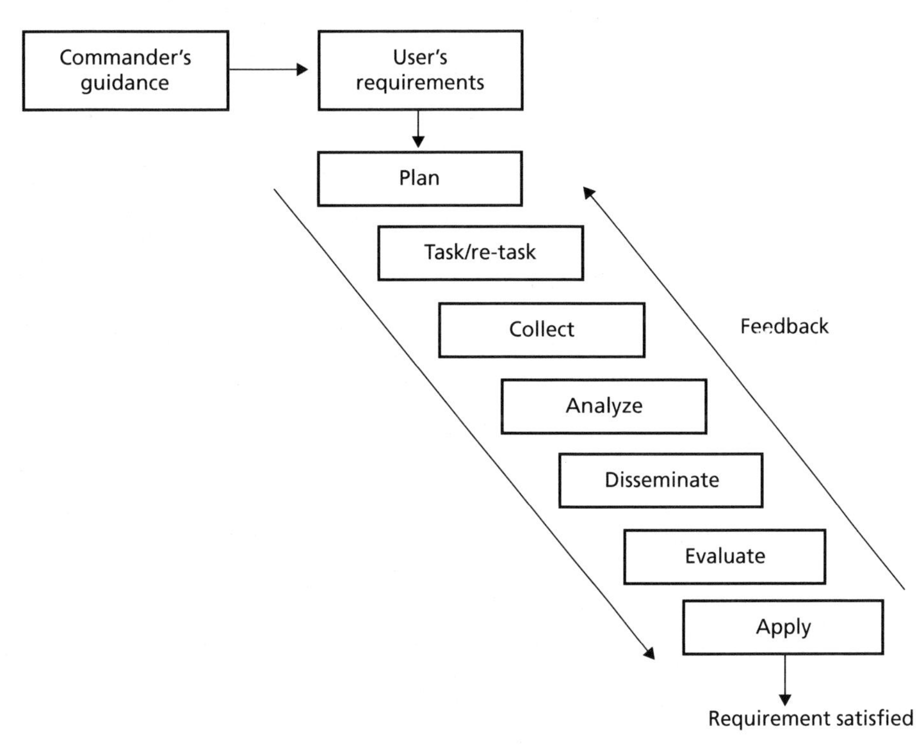

SOURCE: AFOTTP 3–3.6, p.15.
RAND *TR459-2.2*

---

[4] The Air Force is currently revising its primary ISR doctrine, AFOTTP 3-3.6. We believe this is a perfect opportunity to provide clear guidance on ISR assessment and prescribe best practices for use within the Air Force and in other services.

[5] Although intelligence, surveillance, and reconnaissance are three distinct tasks and mission areas, we use the term "ISR system" to include the platforms and processes enabling intelligence collection and dissemination under the control of a Joint Forces Air Component commander or joint forces commander.

Air Force doctrine gives little detail on accomplishing this evaluation, offering but a single paragraph:

> After receiving the ISR products, the user evaluates the product to ensure that it satisfies the requirement. The user then provides feedback to ISR planners, collection managers, and analysts to ensure that the process continues to satisfy the requirement. The user finishes the evaluation step by deciding on appropriate application for the ISR products (AFOTTP 3-3.6, p. 23).

In contrast, joint doctrine offers several references to ISR assessment in Joint Pubs 2-0, 2-01, and 3-30. In a section dedicated to evaluation and feedback, Joint Pub 2-0 (p. II-14) states:

> During the evaluation and feedback phase, intelligence personnel at all levels assess how well each phase of the intelligence cycle is being performed. Commanders and operational staff elements must provide feedback. When areas are identified that need improvement, the necessary changes are made.
>
> Evaluation and feedback are continuously performed during every other phase of the intelligence cycle. Personnel involved in different phases coordinate and cooperate to identify if transitions from one phase to another require improvements. Individual intelligence operators aggressively seek to improve their own performance and the performance of the processes in which they participate.

What remains unstated in these prescriptions is the reasoning behind the need for accurate and timely ISR assessment beyond simply improving the performance of the system. Without this strong motivation, it is easy to let ISR assessment become a low priority during high tempo operations.

Since the supply of ISR resources tends to be more limited than the demand, making the best possible use of the capacity available is critical. The assessment and feedback process makes it possible for ISR managers to determine what ISR systems are being used most effectively and which are not contributing or performing as expected. This evaluation should range from a strategic view of how effectively ISR is supporting campaign objectives down to a tactical view of how particular sensors or communications links are being utilized. Only by assessing a broad range of performance attributes can these types of appraisals be performed and followed up with improved approaches. And when the feedback loop is closed and lessons learned are incorporated in future planning, the overall effectiveness of the ISR system will be improved.

Improved Utilization of ISR Systems

It is self-evident that there is no way to determine how the intelligence process is functioning unless its performance is measured. Furthermore, a well-constructed assessment system should provide managers direction as to the areas needing improvement. It is important here to distinguish between evaluating the information collected, clearly a vital part of providing intelligence to users, and evaluating whether the intelligence system itself is functioning optimally.

As Figure 2.2 illustrates, the intelligence process begins with the commander's guidance leading to requirements, typically represented as a list of PIRs and ISR tasks, each defined in more detail by a set of EEIs. As the military operation evolves, these PIRs and EEIs evolve as well, typically being tied to specific campaign phases and decision points within those phases. As EEIs and then PIRs are satisfied, ISR resources can be freed up and moved on to other tasks. User feedback and continual assessment of ISR performance are critical so that commanders and ISR managers can determine whether progress is being made in fulfilling the PIRs. Perhaps even more important, this assessment plays a vital role in determining whether the PIRs and EEIs being used are even appropriate. If the assessment and feedback mechanisms are not functioning correctly, ISR resources can remain tied up attempting to address requirements that, in reality, have already been satisfied. Similarly, lack of assessment can lead to continuing fruitless collections if, for example, a particular sensor is not appropriate, targets are not where expected, or reports are not being disseminated. It is critical to make the ISR process a "closed-loop" system that can continually self-correct. Assessment and feedback are the links from the intelligence outputs back to the collection managers.

Improved Campaign Execution

As mentioned above, each phase of the campaign has intelligence requirements associated with it. To take a simple example, suppose that a commander wishes to gain air superiority before launching an invasion with ground forces. To gain air superiority, the commander may want all of the surface-to-air missiles (SAMs) found and destroyed. Thus, finding air defenses might be a PIR for the first phase of the campaign. Note the order of prerequisites here: Finding the SAMs is a necessary step in accomplishing the operational objective of destroying them, which in turn is a prerequisite of the strategic objective of air superiority. Poor performance in the ISR system not only slows down the accomplishment of intelligence tasks, but it can also be a limiting factor in the progression of the campaign itself.

However, even this understates the importance of appropriately assessing ISR. Before the campaign can progress, not only must the objectives be accomplished, i.e., the SAMs found and destroyed, but that accomplishment must be confirmed through combat and operational assessments. In the simplest case, BDA must be performed on the SAM sites that were attacked. Since the ISR system will be the source of the information supporting combat assessment, it is again a potential limiting factor. If the ISR system is underperforming, all of these processes will be constrained.

Intelligence gathered by surveillance or reconnaissance systems can also be the limiting factor in a tactical sense. Individual missions may need updated targeting information or target identification before execution. They are dependent on collections to accomplish the mission and may not be able to fly without it. Similarly, if timely Combat Assessment (CA) is not conducted, the mission may have to be repeated or its accomplishment will remain in doubt.

Note that the ISR system *may* be performing correctly, and all of these tactical-, operational-, and strategic-level intelligence tasks *might* be getting accomplished without the need for any assessment of ISR performance. If a minimal level of feedback is received revealing whether PIRs are satisfied on time, time-critical targets are being found, and requestors are not complaining, it is tempting to assume the ISR system is performing well. However, unless introspection of the intelligence enterprise is taking place to measure the efficacy and effectiveness of internal processes and products, and unless detailed feedback from all the external consumers of the intelligence is being gathered, incorporated, and acted upon, there is no way to know whether the ISR system is accomplishing all that is possible. If it is not, it is likely that the entire campaign is being constrained.

Joint Pub 2-0 begins by describing the intelligence cycle (see Figure 2.3) and then proceeds to discuss each step in the cycle in more detail.[6] Whereas the Air Force depiction in Figure 2.2 highlights evaluation and feedback as separate discrete steps, Joint Pub 2-0 highlights them as continuous processes that are involved in every step of the cycle.

Joint Pub 2-0 lays out the tasks and responsibilities of the J-2 and his or her intelligence staff during each phase of the intelligence cycle. One task highlighted during the planning phase is to

> [m]onitor the results of the other phases of the intelligence cycle to determine if PIRs and information requirements are being satisfied. The effectiveness of the collection plan in meeting the JFC's requirements is continually assessed by the command's collection managers (Joint Pub 2-0, p. II-7).

Figure 2.3
Joint Pub 2-0 Description of the Intelligence Cycle

SOURCE: Joint Pub 2-0, p. II-1.
RAND *TR459-2.3*

[6] This publication is currently under revision and in the coordination process. A draft version dated July 14, 2006, exists. We refer to the latest approved version.

A section of the document is devoted to the evaluation and feedback phase; a key quote was given earlier in this chapter.[7] In addition, this section defines seven attributes of intelligence that are to be "qualitative objectives" and "standards against which intelligence activities and products are to be measured." The following are the desired attributes: timely, accurate, usable, complete, relevant, objective, and available. The document goes on to note, "failure to achieve any one of the attributes may contribute to a failure of operations."

Joint Pub 2-01 offers a bit more detail on the intelligence process in general, and ISR assessment in particular, offering several references throughout. Most, however, are repetitive of the guidance offered in Joint Pub 2-0, such as:

> Intelligence operations, activities and products are continuously evaluated. Based on these evaluations and the resulting feedback, remedial actions should be initiated, as required, to improve the performance of intelligence operations and the overall functioning of the intelligence process (Joint Pub 2-01, p. xvii).

These types of statements clearly set a requirement for obtaining feedback and performing ISR assessments, but they are not much help in determining how to carry them out. In addition, Joint Pub 2-01 offers yet another depiction of the intelligence process, shown in Figure 2.4. This document, the most recently updated of those discussed thus far, restores evaluation and feedback as a discrete task but once again connects it with all of the other steps in the process.

As with Joint Pub 2-0, the emphasis in Joint Pub 2-01 is to examine each step in the intelligence process and describe how it is to be performed. As expected, most of the guidance on evaluation and feedback is concentrated in a dedicated section, which begins as follows:

> All intelligence operations are interrelated and the success or failure of one operation will impact the rest of the intelligence process. It is imperative that intelligence personnel and consumers at all levels honestly evaluate and provide immediate feedback throughout the intelligence process on how well the various intelligence operations perform to meet the commander's intelligence requirements (Joint Pub 2-01, p. III-56).

Although this again does not provide much direction on accomplishing the evaluation, it does furnish two things. First, it is very clear on the need for assessment, and second, it sets out a clear directive for producers and consumers of intelligence to provide feedback. Despite the fact that this responsibility is reiterated several times in that publication, as we shall see later in other documentation, this feedback often does not appear to be reaching those who need it.

Joint Pub 2-01 also revisits the attributes of good intelligence, providing eight instead of seven of them. The one added is "anticipatory." The document concludes with an appendix laying out the execution responsibilities for each step in the intelligence process. For the evaluation and feedback step, the responsibilities are shown in Table 2.1. Obviously there is not much detail here, although again the importance of providing feedback at all levels is highlighted.[8]

---

[7] The document, however, does not specify who should carry out this assessment.

[8] Actually, feedback from requestors and "feed forward" to requestors are being recommended here.

Figure 2.4
Joint Pub 2-01 Description of the Intelligence Process

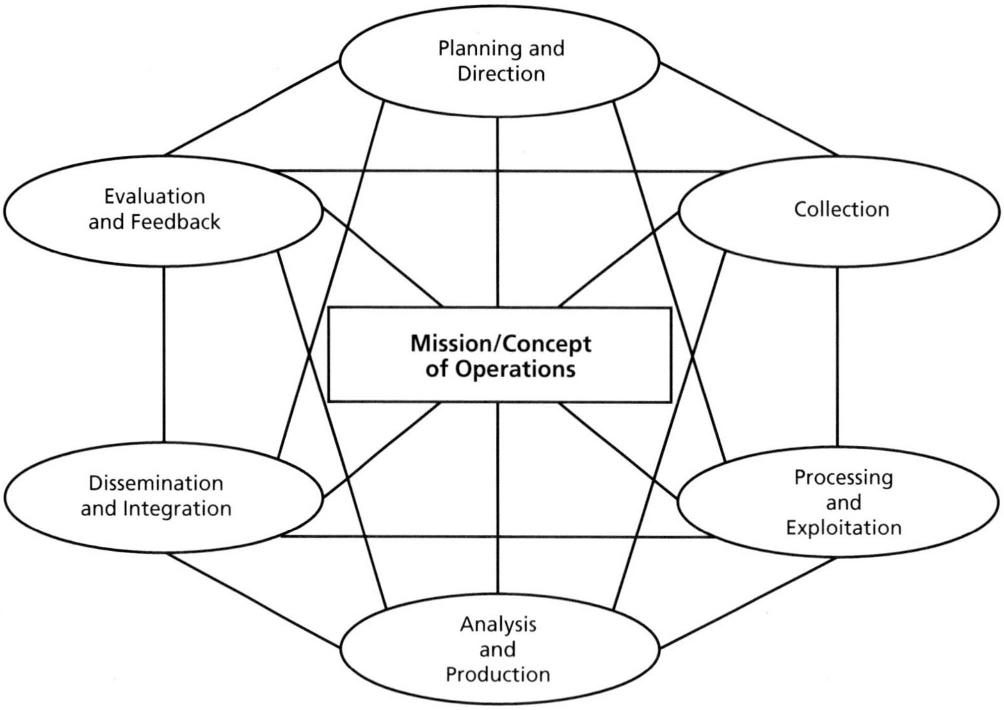

SOURCE: Joint Pub 2-01, p. III-1.
RAND *TR459-2.4*

Table 2.1
Joint Pub 2-01 Responsibilities for Evaluation and Feedback

Organization	Tasks Responsible for
Joint Staff J-2/Defense Intelligence Agency	Provide customer satisfaction and feedback
Combatant command J-2	Recommend improvements for "push/pull" products Provide feedback to requestors
Subordinate joint force J-2	Provide feedback to requestors
Subordinate joint forces components	Provide feedback on all products requested/pushed
Military services	Provide feedback on all products requested/pushed

SOURCE: Joint Pub 2-01, p. H-6.

Air Force CONOPSs and TTP

In response to the joint and service doctrinal shortfalls just highlighted, the Air Force has made a series of efforts over the past several years to provide direction on ISR assessment through other guidance. Some are more helpful than others. For instance, in Air Force Instruction 13-1AOC, the organization of the ISRD in the AOC is laid out and tasks assigned to various cells within it. The ISR management team is directed to "monitor and evaluate the ISR strategy for effectiveness in meeting overall ISR requirements, JFC/JFACC [Joint Forces Air Component Command] PIR, and supporting JFC/JFACC strategy and plans" (Air Force Instruction 13-1AOC, p. 62). Unfortunately, this is the only reference to ISR assessment in the instruction. ISR personnel assigned to the Strategy Division are mentioned as offering support, but primarily for traditional intelligence briefings and operational assessment. On the other hand, U.S. Air Force, Air Force Space Command, 2003, does not mention ISR assessment at all.

The Air Combat Command (ACC) has a significant responsibility for training and tactics development in the intelligence field. For instance, it offers U.S. Air Force, Air Combat Command, 2004, Volume VII of which relates the collection-management process to combat assessment. Collection management here is broken into seven steps, the first of which is requirements and the last is feedback. In a discussion of collection requirements management, one task assigned is "evaluating report feedback/user satisfaction." For the feedback step itself:

> This final step is important for two reasons. First, feedback provides metrics to determine the rate at which unit requests are satisfied. Second, and most importantly, customer feedback identifies problem areas in the collection process and facilitates their solutions (U.S. Air Force, Air Combat Command, 2004, p. 13-8).

This is not very different from what other guidance offers. The document concludes with a sample agenda for the JCMB meeting. One item, labeled "previous collection success/failure" does give two examples of assessment metrics: percentage of successful BDA collection and feedback to requestors regarding non-ATO supported requirements. As we will soon see, this is the first of many examples of measures of effectiveness to be offered.

The most helpful of all current guidance is AFOTTP 2-3.2. This document lays out the organization of the AOC and discusses the tasks and responsibilities of each division and cell within it. The two most relevant chapters for this discussion concern the Strategy Division and the ISRD. These two chapters contain about 15 references to ISR evaluation and assessment.

The ISRD is expected to assign several personnel to the Strategy Division to provide direct intelligence support. One responsibility is for ISR operational assessment:

> The ISR Operations Strategist assesses the operational effectiveness and efficiency of JFACC-assigned ISR assets, strategy, and plans in meeting the intelligence requirements of the JFC, JFACC and other components. ISR operations assessment feeds the overall operational assessment (AFOTTP 2-3.2, p. 3-3).

This statement is critical because it squarely places the accountability for ISR assessment on a particular group and also links this evaluation to the larger operational assessment process. This link is useful because the operational assessment process is well defined within the AOC and JTF and has a team of personnel dedicated to it. Furthermore, it puts ISR assessment on an equal organizational footing with combat assessment—a more traditional and familiar task. It is interesting to note that this is an obvious link to make; yet Air Force and joint doctrine have not done so.

The specific support to be offered to the Operational Assessment Team (OAT) by the ISR operations strategist is detailed later in AFOTTP 2-3.2, specifically in section 3.5.4.6.3. The crux of the ISR assessment problem is summarized here very neatly:

> ISR operations assessment involves the combination of a number of potential factors. The easiest to measure are mission operational factors, such as platform, sensor, crew, or link issues and target deck satisfaction. However, the most difficult aspect of ISR operations assessment is the subjective aspect concerning the answering of PIRs and completion of collection tasks based on MOEs [measures of effectiveness]. This involves the ISR process from strategy to task to the actual sensor tasking and PED [processing, exploitation, and dissemination] and requires feedback from the analysts/customers (AFOTTP 2-3.2, p. 3-69).

This quote highlights the most often reported complaint from intelligence producers and consumers alike—too much emphasis on "bean counting" of sorties flown, hours spent observing, and percentage of targets collected and too little on whether the ISR effort is actually supporting the commander's objectives. The reason for this emphasis, of course, is that the former is fairly easy to calculate and the latter quite difficult to determine, especially given the time pressures of an ongoing campaign. This statement also highlights the existence of ISR tasks beyond the PIRs and the use of measures of effectiveness (MOEs) tied to those tasks and PIRs. These MOEs, which are discussed in more detail later, are obviously critical to evaluating whether a particular PIR or ISR task has actually been achieved. ISR objectives must be written in a way that can be measured; otherwise, their satisfaction will always be in doubt.

The AOC TTP goes on to list the necessary inputs for the ISR assessment process, including a set of questions to be answered for each ISR collection discipline, such as imagery intelligence (IMINT) and signals intelligence (SIGINT). The key items in the list are updated daily and include the PIRs from the JFC, JFACC, and components; the Joint Air Operations Plan (JAOP); the AOD;[9] the RSTA annex; operational results data;[10] feedback from other Joint Air Operations Center (JAOC) elements and staffs on whether their PIRs were being satisfied; PIR-related measures of effectiveness; MOEs for other ISR tasks; and an assessment of the various ISR collection capabilities.

The list of inputs again mentions the existence of ISR tasks separate from the PIRs. In fact, there is an explicit reference:

> Note: PIRs represent only a subset of the overall ISR objectives. Collection tasks/MOEs will be developed for PIRs but will also be developed for other JFACC ISR objectives in addition to the PIRs (AFOTTP 2-3.2, p. 3-70).

Presumably these additional collection tasks would be part of the ISR operations strategy detailed in the AOD, and as noted here, would have their own MOEs to enable assessment. The relationship between the PIRs and these "other" tasks is not detailed, however. It is not clear from this guidance what these other tasks are or how their priority is to be compared with that of the PIRs. They must be either prestrike or BDA collections supporting the targeting process; lower-priority tasks that are "nice to have" if collection capacity is left over after satisfying the PIRs; or they are additional tasks, not reflected in the PIRs, that are necessary to satisfy the commander's objectives. In the latter case, revision or expansion of the PIRs would seem to be in order, rather than the creation of a new, seemingly independent, set of ISR tasks.

---

[9] The AOD includes the ISR operations strategy, tasks, and measures of effectiveness.

[10] These data include traditional quantitative measures such as ISR platform and PED performance, percentage of planned collections that were accomplished, timeliness, communication rates, and ad hoc success rates.

According to AFOTTP 2-3.2, the assessment of the various ISR collection capabilities should be conducted by answering a series of questions about each intelligence capability. Although the questions vary by specific intelligence type, they are generally of the form:

- Does the collection deck accurately reflect priorities and guidance?
- Are we tasking the right sensor against the right target?
- What is the effectiveness of our airborne tracks? Could they be moved to improve effectiveness?
- Are ad hoc requests being given the proper priority? Are they preventing the satisfaction of PIRs? Should more flexibility be built into the preplanned collection deck?
- Are we accurately deconflicting resources?
- Are collection elements using the most appropriate reporting vehicles in support of requirements? Are the vehicles and timelines sufficient to meet tasked objectives? Are real-time dissemination techniques conveying the desired depth of information?
- Are we scheduling and placing our airborne assets to coincide with windows of opportunity for collection, given current IPB [intelligence preparation of the battlespace]? Are theater assets located for optimum geolocation and collaboration?

In addition to the feedback elements mentioned above, these questions attempt to get at the qualitative aspects of ISR assessment and are a mix of questions to determine whether the ISR system is performing as expected and, if not, how it should be corrected. One could argue, however, that essential questions about whether ISR efforts are producing the desired effects are still missing. Some of these questions are fairly straightforward and can be answered with a small piece of analysis, but others will be very difficult to address with a simple yes or no. Obviously, ISR staff in the Strategy Division cannot answer them all themselves; they must rely on the ISRD to support the ISR assessment portion of operational assessment. Ultimately, these types of questions might be made most useful simply by providing a series of reference points and reminders to the J-2, ISR and Strategy Division staffs during the ISR planning process.

Chapter 6 of AFOTTP 2-3.2 details the organization and tasking of the ISRD. One cell of the division is the Analysis, Correlation, and Fusion Team (ACF), with the responsibility for conducting IPB and supporting predictive battlespace awareness, which is detailed as a cycle, shown in Figure 2.5, and is quite similar to other depictions of the intelligence cycle shown so far in this report.

The process shown here, of course, does depict not only a single day of the cycle, but also activities taking place both before and during a conflict. ISR assessment is again highlighted here, not as a part of operational assessment but as a separate assessment task. This formulation does not appear to be aligned with the description given earlier in AFOTTP 2-3.2, Chapter 3, in which ISR assessment was an integral part of operational assessment.

Other descriptions of the ISRD include a reiteration of the personnel assigned to the Strategy Division and their roles in ISR assessment. Although all the other teams are given a responsibility for assessment,[11] the ISR operations strategists in the Strategy Division are clearly designated as the primary evaluators of ISR. More specific steps in the assessment process are given in AFOTTP 2-3.2 (p. 6-99), specifically:

1. Coordinate with relevant entities—component collection managers/analysts/liaison officers (LNOs), the ISR Operations Team, the ACF, the Processing, Exploitation, and Dissemi-

[11] Particularly, responsibility is given to the PED Cell. See, for example, AFOTTP 2-3.2, pp. 6-150 through 6-152.

Figure 2.5
Depiction of Predictive Battlespace Awareness in AFOTTP 2-3.2

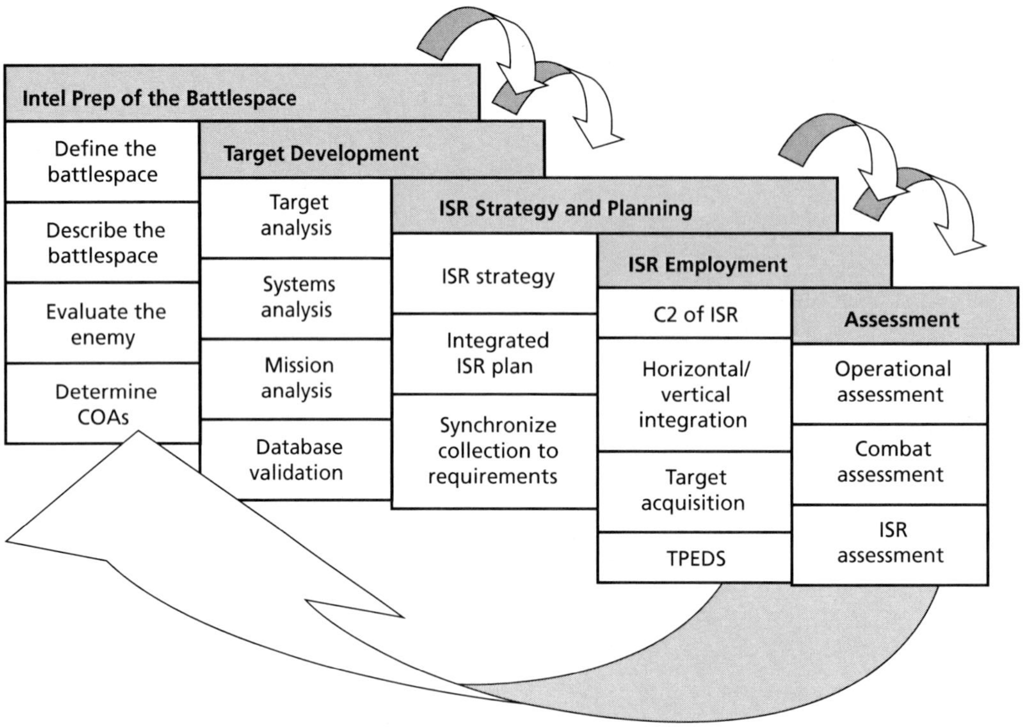

SOURCE: AFOTTP 2-3.2, p. 6-12.
NOTES: Intel Prep = intelligence preparation; COAs = courses of action; C2 = command and control; TPEDs = tasking, processing, exploitation, and dissemination.
RAND *TR459-2.5*

nation (PED) Management Team, the ISR operations duty officer, platform/PED LNOs, etc.—for obtaining the required data to make the ISR operations assessments.

2. Develop, in coordination with the intelligence analyst and target planner in the Strategy Division and other components, MOEs for determining whether PIRs/ISR tasks are being fulfilled.
3. Develop an ISR input to the operational assessment plan to evaluate JAOP planning and execution.
4. Monitor and evaluate the ISR strategy for effectiveness in meeting overall ISR requirements, JFC/JFACC PIRs, and supporting JFC/JFACC strategy and plans.
5. Assess and report on the ISR sections of the JAOP and AOD and on RSTA annex effectiveness in terms of objective and task accomplishment, adherence or divergence from the established plan, and optimum use of available resources (e.g., ISR platforms, sensors, PED assets).
6. Produce ISR operations assessment documentation (e.g., briefings, reports) as required.

As with descriptions of the assessment process in other guidance documents, this "game plan" is fairly clear about what needs to be done, but not how to do it. The set of inputs and questions to the assessment process given in Chapter 3 of AFOTTP 2-3.2 (see discussion above) is more specific in how to accomplish the steps laid out here.

Moving now to the lower levels of service guidance, the Air Combat Command recently sponsored a conference on ISR assessment that led to drafts of an ISR assessment functional concept (U.S. Air Force, Air Combat Command, 2005b) and an ISR assessment CONOPS (U.S. Air Force, Air Combat Command, 2005a). Since these documents are similar and the functional concept is more recent and appears more complete, we will concentrate on it.

The key addition offered in those drafts to the guidance already examined is the separation of ISR assessment into three levels: tactical, operational, and strategic. They are defined as:

> *Tactical ISR Assessment.* Determines if specific ISR missions provided desired intelligence based on assigned tactical tasks. Key question to answer: Were the individual tasks assigned to ISR operators/assets performed successfully?
>
> *Operational ISR Assessment.* Determines if the C/JFACC ISR strategy and the aggregate of ISR operations have produced the desired effects based on stated objectives. Key question to answer: Are ISR operations adequately supporting operational objectives?
>
> *Strategic ISR Assessment.* Determines if the ISR strategy and resources are appropriate to achieve theater/campaign objectives. Key question to answer: Are ISR operations satisfying theater requirements? (U.S. Air Force, Air Combat Command, 2005b, p. 4.)

Given the set of metrics that each of these assessment tiers is matched with, there is a natural association between the tactical level and quantitative assessment, and the strategic level and qualitative feedback. The operational-level assessment would utilize a mix of both. It is also interesting to note that none of these definitions refers to PIRs, only to the three levels of objectives typically present in a strategies-to-tasks framework. In fact, a few pages later the document states:

> Combatant Command and component level/supported command PIRs may not be synchronized, and therefore, not necessarily reflect requirements that support the Joint Force Commander's overall campaign. Poorly synchronized objectives and PIRs make overall ISR Operations difficult to assess and have a negative impact on subsequent steps in the process (U.S. Air Force, Air Combat Command, 2005b, p. 7).

The issue raised here could be addressed by ensuring that PIRs and campaign objectives are aligned. Unfortunately, this task is typically out of the purview of the JFACC's intelligence staff, although not of the Strategy Division. Another approach could focus on accomplishing the commander's objectives as the sole measure of ISR success and view the PIRs only as a tool for translating objectives into collections.

For each of the three levels of assessment, the functional concept document goes on to describe the components and data sets that need to be collected. The general hierarchy, with some detail removed, is as follows (U.S. Air Force, Air Combat Command, 2005b, p. 10):

I. Tactical
    a. Execution
        i. Data link architecture
        ii. Aircraft operations

iii. Processing
iv. Exploitation
v. Dissemination
vi. Dynamic ad hoc tasking
vii. Mission planning

b. Planning
i. EEIs
ii. Tasking
iii. Feedback process

II. Operational
a. Execution
b. Planning
i. Collection strategy
ii. EEIs
iii. Tasking
iv. Feedback process
c. Strategy
i. Commander's objectives and PIRs
ii. Apportionment

III. Strategic
a. Commander's objectives and PIRs
b. Apportionment.

While this framework is mostly clear and organized, providing a good set of measures to gather, again we note that the more quantitative tactical level is described in much more detail than the higher, and presumably more important, operational and strategic levels. Perhaps this lack of detail is due to the ease with which hours, collections, and products can be counted versus the difficulty in assessing whether a broad set of multiple types of intelligence collections have really satisfied a PIR.

The ISR assessment functional concept then goes on to match a specific organization with the piece of the ISR assessment for which it is responsible. As expected, all of the cells in the ISRD have an assessment responsibility, with the PED Management Team performing tactical assessment and operational and strategic assessments being done by the Strategy Division OAT. The Strategy Division is also tasked with providing the assessment report and recommendations for improvement to the JFACC. These responsibilities are similar to that detailed in AFOTTP 2-3.2, although that document does not divide ISR assessment into the three levels described in U.S. Air Force, Air Force Doctrine Center, forthcoming.

Command-Specific Techniques

To implement all of the guidance discussed to this point, it is up to the personnel actually assigned to the JFC J-2 staff, AOC, and component intelligence staffs to devise specific, executable approaches. As a short survey of some of these techniques, we have reviewed several documents and briefings from Central Command Air Forces (CENTAF),[12] Pacific Air Forces (PACAF)[13] and U.S. Air Forces in Europe (USAFE).[14] We will discuss each of these in turn, by organization.

CENTAF appears to have the most advanced techniques of these three component commands, with a formal ISR assessment CONOPS and TTP in draft form. Given the ongoing warfighting of the command, this advanced status should not be surprising. The CONOPS document offers a formal definition of ISR assessment as:

> Evaluating the TCPED [tasking, collection, processing, exploitation, and dissemination] process end-to-end to ensure all CFACC [combined force air component commander] controlled ISR assets are optimally employed and the derived information is accessible and relevant to warfighters, planners and decision makers (U.S. Central Command Air Forces, 2005a, p. 1).

As a means to this end, the CONOPS document discusses two main courses of action: expanding traditional quantitative PED statistics and merging those statistics with qualitative feedback. According to CENTAF, the ultimate goal of ISR assessment should be to answer such questions as the following (U.S. Central Command Air Forces, 2005a, p. 2):

- Has the user been educated on platform capabilities, to include cross-cue, in order to understand the available effects?
- Were EEIs properly constructed/written to get the desired effect?
- Was the appropriate sensor tasked to achieve the customer's requested effect?
- Did the tasked sensor achieve the requested ISR effect?
- Does the user know where to access finished products?
- Was the user satisfied with the product they received?
- What was the effect of ISR products on user operations?
- Was the tasked sensor optimally employed to achieve the effect?

It is interesting to compare this list of questions with the list above from AFOTTP 2-3.2. They are fairly similar, although the list here concentrates more on effects and the user's opinion of success, rather than on an internally focused assessment of the TTP document. There is also no reference here to the three levels of assessment put forth in the draft functional concept from ACC.

The remainder of the CENTAF draft CONOPS contains a discussion of the tasking, collection and processing, and exploitation and dissemination steps of the intelligence cycle and gives suggested feedback items and quantitative measures to evaluate success in each step. Most of the tasking items revolve around evaluating requirements and gathering feedback on shortfalls from customers and flying units. The collection and processing step contains most of the quantitative measures, although collection "satisfaction" is one of them. This metric, defined for IMINT as, "Did the customer get the image they needed?" is obviously subjective in nature, but it does lead to much lower success rates than the more typical, "Was everything collected that was planned for?" The exploitation and dissemination evaluation is based mainly

---

[12] U.S. Central Command Air Forces, undated[a], undated[b], 2005a, and 2005b.

[13] 26 Air Intelligence Squadron, undated; and Zwicker, 2005.

[14] Brown and Pearson, undated.

on feedback with "better user education" mentioned as the most obvious solution. Although not mentioned in the document, there are quantitative metrics that could be added here as well, such as "percentage of posted reports that were viewed" or "average time spent by users searching for a report."

The two CENTAF briefings on ISR assessment both mention the difficulties raised by relatively simple quantitative metrics. To quote one, "Metrics don't tell [the] qualitative story . . . leads to bean counting." Obviously qualitative shortfalls are being observed that are not being captured by the quantitative metrics. However, the briefing continues, "Customer feedback is slim, often anecdotal and non-specific" (U.S. Central Command Air Forces, undated[b], p. xx). As a result of this quandary, the command has made particular efforts to collect a broad set of metrics and to maintain a database of preliminary and postmission summaries for its own analytic use. In addition to these mission reports, the command also collects feedback using Web pages, email forms, phone calls, and face-to-face meetings from the Combined Air and Space Operations Center (including platform LNOs, ISRD teams, and the leadership), at the supported components, at ISR platform units, and at intelligence reach-back agencies. All this information enables analyses, such as sensor and platform comparisons.

The two briefings from PACAF/INXP and the 26th Air Intelligence Squadron contain somewhat similar viewpoints, but they appear to be more oriented toward implementing existing joint and Air Force guidance than developing new CONOPSs. The overall process is broken into three parts here: operational effectiveness, which is generally seen as quantitative; mission effectiveness, which is qualitative and more difficult; and strategic assessment (Zwicker, 2005, p. 3). This three-level construct is similar to that defined in the ACC functional concept. The Strategy Division in the AOC is highlighted as the Office of Coordinating Responsibility, but the "bottom line" from the 26th Air Intelligence Squadron (AIS) briefing is: "Need a continuous ISR Assessment process that feeds back to the Strategy Division" [emphasis in original] (26 AIS, undated, p. 3).

These two briefings also highlight the use of a strategies to ISR-tasks matrix generated by ISR strategists in the Strategy Division. This matrix includes measures of effectiveness for objectives and measures of performance for tasks. Although this is similar to the MOE framework discussed in AFOTTP 2-3.2, no reference is given here to how PIRs and EEIs are treated in this strategies-to-tasks architecture. Recall that the TTP document (AFOTTP 2-3.2) treated PIRs and ISR tasks as two separate sets of requirements. However, when later in the AIS briefing an ISR mission assessment tool is discussed, EEI and PIR satisfaction appear to be the main objectives (26 AIS, undated, p. 10). In this tool, a red-yellow-green color spectrum is used, with red indicating that the collection requirements were not satisfied and EEIs were not answered, yellow indicating that most of the collections and most of the EEIs were answered, and green indicating that all of the collections and EEIs were satisfied and the PIRs were answered. It is interesting to note that in order to gain a higher "score" (greener color), more qualitative and feedback-based assessment must be performed. Purely quantitative assessment could, at best, result in a yellow "score."

Three main sets of output of the assessment process are also discussed: the commander's critical information requirements, PIRs, and EEIs; ISR "reattack" recommendations; and revised inputs (26 AIS, undated, p. 11). The first addresses whether the commander's questions are being answered and also whether the right questions are being asked. This point is important and should not be neglected. Even if the ISR system is perfectly addressing the commander's requirements, if those requirements are not the correct ones to support the overall objectives of the campaign, the ISR system should not be considered effective. The second output concerns mainly optimum asset employment as governed by the ISR strategy, and the last examines the MOEs and measures of performance defined in the strategies to ISR-tasks framework.

The final descriptive document we examine here is from the USAFE AOC ISRD staff (32nd AIS). This briefing shows a single day's ISR planning process and highlights various assessment-related activities, with an emphasis on ISR effects. The strategies to ISR-tasks framework is emphasized again, with several real-world examples given (Brown and Pearson, undated, p. 14). Although PIRs are mentioned as being linked to the framework and example PIRs are included, no discussion of a link is given. This problem occurs in several of the documents we reviewed. The briefing highlights how desired ISR effects are attached to the tactical ISR and strike-support tasks, which in turn are then synchronized. The effects and resulting timelines are then disseminated in the daily RSTA document and are used to evaluate the effectiveness of the day's ISR operations. As an example, if the desired effect is confirmation of a target kill by a certain time, and if the collection cannot discern the target's status or if it is received too late, then the collection should not be considered a success, even if it was collected.

The Brown and Pearson document is the first appearance of ISR assessments looking for ISR effects. The general idea there is to move away from assessing whether a particular target was collected and move to assessing whether the desired intelligence effect was achieved. This outcome is similar to effects-based targeting, which, instead of looking at how many targets are struck, focuses on striking the right ones at the right time to achieve the desired effect on the adversary. One outcome of assessing ISR effects is a reduction in the reliance on bean counting of collections. Properly integrating this approach into ISR operations requires moving beyond the traditional PIR-EEI-observable method of describing requirements to one that specifically describes every ISR effect desired.

The roles of the senior intelligence duty officer (SIDO) in the AOC and the ACF senior analyst are highlighted as the keys to the assessment process in this document (Brown and Pearson, undated, p. 24). The SIDO log provides real-time reporting on whether the desired ISR effects were being achieved, as well as the timeliness of ad hoc collections. Presumably, the effect of ad hoc requests on the planned deck would be noted as well. The ACF analyst addresses the PED process and also the utility of the reporting. Although the primary question being asked is "Did we answer the PIRs?" it is not entirely clear from the process described in Brown and Pearson how the quantitative data and qualitative feedback are merged to answer that question.

CHAPTER THREE

Basing the Course Ahead for ISR on Experience

As part of this study, we undertook a thorough review of classified and unclassified literature and conducted personal interviews with key players in Air Force intelligence positions from Operation Enduring Freedom (OEF) and Operation Iraqi Freedom (OIF). We also participated in ACC-sponsored conferences and headquarters exercises at PACOM to better understand how lessons learned from the two recent operations were being integrated into current ISR practices.

Although issues have been raised from OEF and OIF, ranging from the need to better integrate human intelligence to a call for revamping the BDA process, we focus here on three main topics related to this research: organization and integration of the ISRD in the AOC, collection management, and ISR assessment. We now discuss the lessons learned from OEF, OIF, and the PACOM exercises for each of these three topics.

Organization and Integration of the ISRD

One common thread emerging from OEF and OIF is the difficulty in integrating intelligence and operations activities. In a joint headquarters, these functions are doctrinally organized under the J-2 and J-3 directorates, while in the AOC they fall under the purview of the ISRD and Combat Operations Division (COD). In addition to the ISRD, the Strategy Division and Combat Plans Division also support the COD in the AOC by generating the ATO and several other inputs.

Concerns from recent operations revolve around the ability of the ISRD to remain synchronized and integrated with the ATO, particularly with production of the collection plan and the RSTA annex.[1] Although the COD has access to significant ISR capabilities (see Figure 3.1) for use in prosecuting time-critical and time-sensitive targets with the SIDO, the basic priorities and objectives of each day's collection plan are set forth in the RSTA annex and ISR section of the AOD.

The RSTA annex and the ISR section of the AOD are critical to providing guidance to the SIDO and the entire COD regarding the proper allocation of ISR resources. If the documents are not synchronized with the JFC and JFACC's objectives and do not accurately consider the current ATO plan, they will provide little, or incorrect, guidance to the SIDO during ATO execution. The problem is compounded because, as Figure 3.1 illustrates, major

[1] Shlapak, 2006; Johnson, 2004; and author interview with Lt. Col. "Gracie" Matthews and Col. (ret.) Ron Chilcote, Air Force Special Operations Command, Intelligence Directorate, Hurlburt Field, Fla., March 4, 2005.

Figure 3.1
Combat Operations Division Organization in AFOTTP 2-3.2

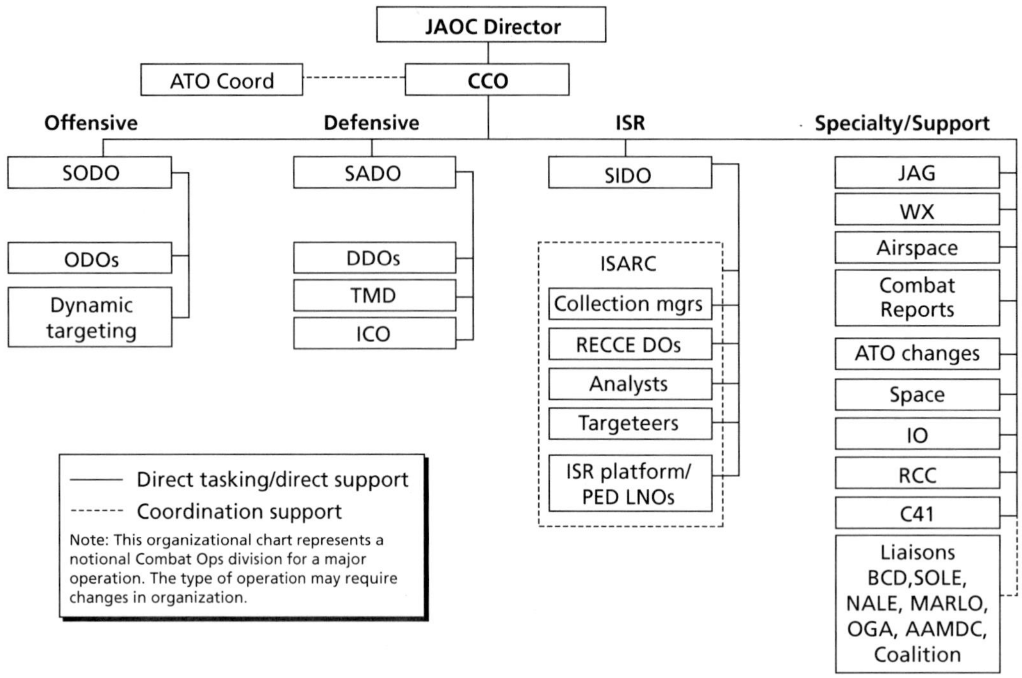

JAOC Director
ATO Coord
CCO
Offensive
Defensive
ISR
Specialty/Support
SODO
ODOs
Dynamic targeting
SADO
DDOs
TMD
ICO
SIDO
ISARC
Collection mgrs
RECCE DOs
Analysts
Targeteers
ISR platform/ PED LNOs
JAG
WX
Airspace
Combat Reports
ATO changes
Space
IO
RCC
C41
Liaisons BCD,SOLE, NALE, MARLO, OGA, AAMDC, Coalition
Direct tasking/direct support
Coordination support
Note: This organizational chart represents a notional Combat Ops division for a major operation. The type of operation may require changes in organization.

SOURCE: AFOTTP 2-3.2, p. 5-2.
NOTES: Coord = coordinator; CCO = commander of combat operations; SODO = senior offensive duty officer; ODOs = offensive duty officers; Tgting = targeting; SADO = senior air defense officer; DDOs = defensive duty officers; TMD = theater missile defense; ICO = interface control officer; Mgrs = managers; RECCE DOs = reconnaissance duty officers; JAG = judge advocate general; WX = weather; IO = intelligence officer; RCC = Rescue Coordination Center; C4I = command, control, communications, computers, and intelligence; BCD = battlefield coordination detachment; SOLE = Special Operations Liaison Element; NALE = Naval and Amphibious Liaison Element; MARLO = marine liaison officer; OGA = other government agency; and AAMDC = Army air and missile defense command.
RAND *TR459-3.1*

components of the ISRD also make up the Intelligence, Surveillance, and Reconnaissance Cell (ISARC). These personnel tend to become continually involved in supporting mission execution and have little time remaining for their planning function (U.S. Central Command Air Forces, 2004).

The synchronization issue raised in OEF and OIF is most likely the result of the organization of the ISRD relative to the rest of the AOC. As shown in Figure 3.2, the ISRD is organized into four functional teams: ACF, Targeting, Operations, and PED Management. The responsibilities of each team are summarized in AFOTTP 2.3-2 as follows:

- The ACF team conducts dynamic intelligence preparation of the battlespace (IPB) that provides the context for understanding the adversary's intentions and supports the application of Predictive Battlespace Awareness (PBA) [p. 6-5].

- The Targets/Combat Assessment (Tgt/CA) Team coordinates targets and combat assessment functions for the JFACC [p. 6-44].

Figure 3.2
Organization of the ISRD

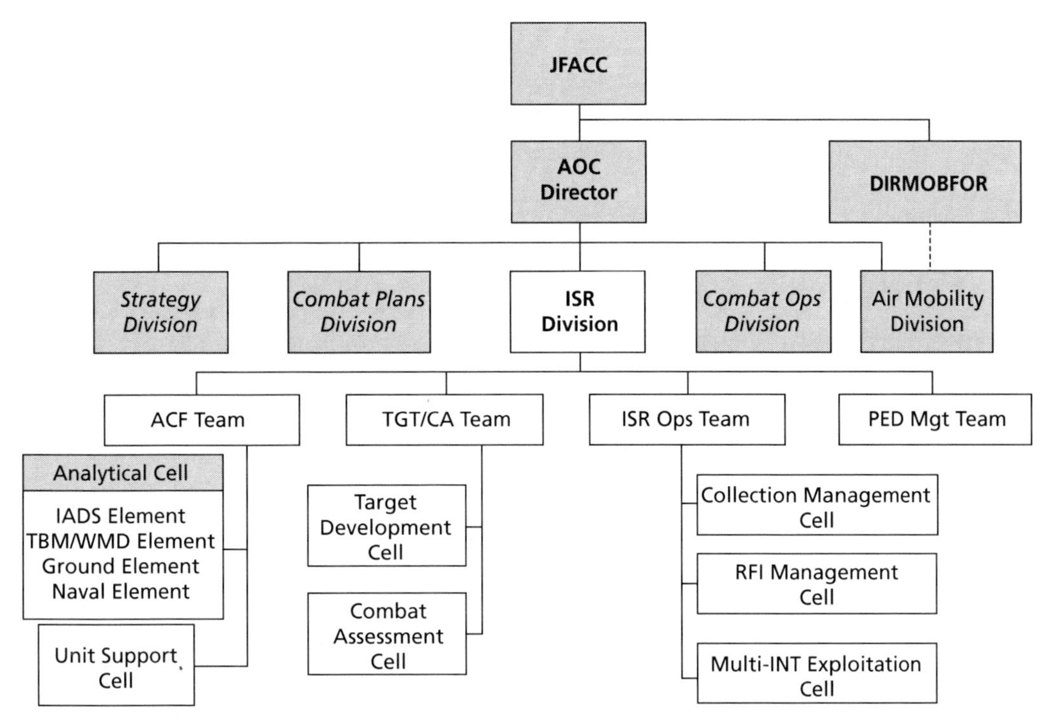

SOURCE: AFOTTP 2-3.2, p. 6-3.
NOTES: DIRMOBFOR = director of mobility forces; Ops = operations; Mgt = management; IADS = Integrated Air Defense System; TBM/WMD = theater ballistic missiles/weapons of mass destruction; PEC3 = political, economic, command, control, and communications; RFI = request for information; and INT = intelligence.
RAND *TR459-3.2*

- ISR operations in the JAOC encompasses the following: guidance and objectives refinement, strategy-to-task development, prioritizing requirements, integrating ISR and combat operations (to include strike operations), developing the ISR platform schedule, collection management, PED (processing, exploitation, and dissemination) management, building collection tasking, tasking the sensors and PED nodes through development of the Reconnaissance, Surveillance, and Target Acquisition (RSTA) annex, execution of the RSTA annex, dynamically adjusting platform/sensor/PED tasking as required during execution, and ISR Operations assessment [p. 6-91].[2]
- The PED Management Team is the ISRD focal point for implementing, coordinating, and maintaining PED support from units/agencies outside the JAOC . . . . The PED Management Team monitors ISR assets and PED mission execution, collects and analyzes TPED [tasking, processing, exploitation, and dissemination]-related metrics, identifies discrepancies in the PED mission, and institutes control measures to correct or improve the PED process [p. 6-147].

What is important to note about these descriptions is that only one team, ISR Operations, is defined as having a responsibility to assist with ISR replanning during execution. Fur-

[2] AFOTTP 2.3-2 lists PED management as a responsibility of both the ISR operations team and the PED management team. This is a potential point of confusion.

thermore, despite the name of this cell, the majority of its tasks involves planning that must occur both before and after the ATO is developed.[3]

Despite this focus on deliberate planning in the ISRD structure and doctrinal responsibilities, the real world of AOC operations in OEF and OIF has shown that there are simply not enough time and personnel to operate in a serial manner. We can no longer expect step 1 to occur before step 2 and to hold off on step 4 until step 3 is complete. During ATO execution, changes occur so quickly in priorities, enemy courses of action, and available resources that ISRD personnel on all four teams inexorably get drawn into supporting the SIDO and the ISARC and, thus, are unable to adequately perform their doctrinal (and trained for) responsibilities. In addition, the time available for these tasks has become compressed as ATO cycles are shortened and "dynamic" ATOs become the norm.

This pressure toward execution and away from planning is not new to the AOC; the targeting and strike communities have faced it for some time. In response, there is a Strategy Division (long-range planning), a Combat Plans Division (to build the ATO), and a Combat Operations Division (to execute the ATO). Rather than organizing functionally, they have become organized according to time horizon, and they are staffed correspondingly.

There are two obvious potential solutions to allow better synchronization between ISR and operations. The first is to organize the ISRD parallel to the rest of the AOC with strategy, plans, and operations teams. The second is to carry the current practice of embedding ISRD personnel in other divisions to its natural conclusion and disband the ISRD totally—putting in its place complete ISR teams in each of the Strategy, Plans, and Operations Divisions. At this point, it is unclear which course is the best to take, and future work is needed.

The first potential solution, to reorganize the ISRD away from functional teams and into teams by time horizon, is probably the simplest and would require the minimum amount of adjustment. The new ISR "strategy team" would be largely made up of collection managers from the current ISR Operations Cell and, as per AFOTTP 2-3.2, would focus on guidance and objectives refinement, strategies-to-tasks development, prioritizing requirements, collection management, and ISR Operations assessment. Similarly, a new ISR plans team would be responsible for integrating ISR and combat operations (to include strike operations), developing the ISR platform schedule, building collection tasking, tasking the sensors and PED nodes through development of the RSTA annex and target development. Finally, ISR Operations would be responsible solely for execution of the RSTA annex, dynamically adjusting platform, sensor, and/or PED tasking, as required during execution, and PED management. This team would also serve as the ISARC and would report to the chief of combat operations through the SIDO. It would have no collection-management or planning responsibilities.

Note that the ACF has not yet been mentioned. This team consists of subject-matter experts in a variety of fields and so would most likely be of greater utility to the ISRD remaining as it is. If broken up to support the three main teams, much duplication of effort and expertise would likely be required. The challenge, of course, would be to prevent these experts from being consumed by ongoing operations, preventing their support to IPB and other planning processes.

The more drastic solution is to break up the ISRD entirely. Already, the SIDO and ISARC report to the Combat Operations Division chief (see Figure 3.1) to support execution, and the

[3] There can be significant differences between doctrine and actual AOC organization in various theaters. We have attempted to illustrate a number of problems we observed in the field using a doctrinal template for reference.

Targeting Team spends much of its time supporting the Joint Effects Team and the master air attack plan in the Combat Plans Division. ISRD personnel are also assigned to support the Strategy Division. This leaves the ISR short- and long-term planners, the ACF specialists, and the PED managers. ISR campaign planning and collection management could probably occur in the Strategy and Combat Plans Division with little disruption and, in fact, would probably streamline coordinated planning, which is now dependent on embedded personnel. Although such a drastic change could make the lines of communication and command much clearer, it would also leave the ACF and the PED Management Team without a natural home. The most obvious solution would be to leave them as is, in a much smaller intelligence division, in which they would serve in a pure support role with no direct short-term planning or execution responsibilities. The ACF would remain focused on IPB and providing subject-matter expertise to the other divisions as necessary, while the PED Management Cell would remain responsible for turning collected data into information to pass on to the consumers. This solution breaks up a unified command into two functions, one for combat operations collections and the other for PED. The two must be well linked and synchronized to effectively support the commanders' objectives.

Collection Management

The second large issue we address that emerged from OEF and OIF was the inability to properly tie collections back to objectives and to use this link to guide collection planning and dynamic retasking.[4] Col. James Poss (2004, slide 2) summarizes the issues with current doctrine and TTPs in his briefing as:

> Lack of effect[s]-based ISR process, [the] tool causes these problems:
>
> - No clear JFC/JFACC-approved ISR prioritized guidance
> - No clear way to identify ISR requirements for each ATO
> - Collection priorities not clearly tied to tactical tasks; JFC's operational objectives
> - No automated collection-management tool to support ISR planning; support limited to ISR execution only
> - Present tools are mainly visualization tools, focused on ISR ops, not plans or strategy.

Current joint doctrine and Air Force doctrine describe the procedures for turning commander's guidance into actionable intelligence collection targets.[5,6] Ideally, in a wartime situa-

---

[4] Bradley, 2004; U.S. Central Command Air Forces, 2003; author interview with Dean Daigle, 612th Air Intelligence Group, Davis-Monthan Air Force Base, Ariz., March 8, 2005; author interview with Lt. Col. "Gracie" Matthews and Col. (ret.) Ron Chilcote (2005); author interview with Lt. Col. Gregory Brodman, Lt. Col. Jenkins, Lt. Col. Ducharme, and MSgt Orf, Shaw AFB, S.C., May 13, 2003; and Poss, 2004.

[5] Primarily, Joint Pub 2-01; Joint Pub 3-55; and AFDD 2-5.2. AFDD 2-5.2 is currently under revision and in the coordination process. A draft version dated November 21, 2005, exists. However, we refer to the latest approved version. AFDD 2-5.2 will be published as AFOTTP 3-3.6.

[6] Any lack of "actionable intelligence" (known information) in a given contingency is not addressed in our analysis.

tion, the JTF J-2 performs this process, although if the JFACC is designated as the supported commander for ISR, the AOC staff can be responsible for a large part of this process. Even if the supported component is not designated for ISR, the air-breathing platforms make up a large portion of the ISR collection capability, and so the AOC staff can have a large influence on the results. Please refer to Chapter Two of this report for details on the collection-tasking process.

The root cause of the problems raised during OEF and OIF seems, in many cases, to be a lack of transparency and a lack of adequate tools. Transparency is key because the process of moving from commander's PIRs (which are typically phrased as broad, often quite vague questions) to a specific collection target involves several staffs, often in disparate locations, and usually requires quite a bit of interpretation on the part of the collection managers to "fill in the blanks" among commander's objectives, PIRs, and collection requirements. If the reasoning process used and resultant steps taken are not obvious to subsequent analysts, the resulting collections can diverge significantly from the original intentions.

The primary tool used in OEF and OIF for collection planning is the Web-based PRISM. It is used to manage the integration of collection requests from a variety of customers and provide broad insight into the collection-management process. Although it provides transparency into the collection deck, PRISM is fairly limited in the detail it provides to users. Collections are linked back to PIRs, but little further detail is provided about how collections support objectives or what effects are being sought by particular requirements.[7] In OEF and OIF situations in which priorities and objectives were constantly changing during the cycle's planning process, the data rapidly became outdated and cluttered with outdated requirements (author interview with Lt. Col. Gregory Brodman, Lt. Col. Jenkins, Lt. Col. Ducharme, and MSgt Orf, 2003; and U.S. Marine Corps, 2003). While this problem could be rectified with more personnel, enhancements to current tools may be a more cost-effective remedy.

As a potential solution for these collection-management difficulties, we discuss later in this chapter enhancing the collection-management process with a strategies-to-tasks and utility framework. By linking collection targets to operational tasks, objectives, and the top-level commander's guidance with relative utility ratings, planning for the daily intelligence collections and real-time retasking for ad hoc ISR targets could be enhanced. If current tools are modified to provide this information, planners will be able to link collection targets to top-level objectives for better decisionmaking and optimization of limited collection assets. Similarly, at the AOC, intelligence officers will be better able to deal with time-sensitive, emerging targets by rapidly comparing the value of collecting an ad hoc collection with the value of collecting on an already planned one.

ISR Assessment

The final lesson learned from OEF and OIF that we address concerns ISR assessment. As with more traditional combat or BDA, an end-to-end assessment of ISR effectiveness and efficiency should be used to monitor and improve daily operations and ensure that limited ISR assets are being utilized to their maximum capacity. In OEF and OIF, however, ISR assessments have generally focused on statistics from the tactical level (sorties flown, percentage of planned

[7] After this work was briefed to senior U.S. Air Force officers, new naming conventions were added to PRISM to tie collection requirements to specific tasks at the operational level.

images collected, etc.).[8] The question of whether the ISR system is helping to satisfy the commander's intent has gone largely unanswered by assessment mechanisms utilized in the field.

A common complaint from OEF and OIF is that there is too much emphasis on bean counting of sorties flown, hours spent observing, and percentage of targets collected and too little on whether the ISR effort is actually supporting the commander's objectives. The reason for this dichotomy, of course, is that the former is fairly easy to calculate and the latter quite difficult to determine, especially given the time pressures of an ongoing campaign. In most cases, easily calculated metrics are not able to be translated directly to achieving a commander's objectives.

An end-to-end assessment process is needed that can improve daily ISR planning and platform employment as well as assure the commander that his or her objectives are being fully supported by ISR assets under his or her control. To enable this, an ISR assessment process must appropriately integrate a large amount of both quantitative data and often-piecemeal qualitative judgments from sources with vastly differing perspectives. Furthermore, this process must use as few personnel hours as possible so that it does not disrupt ongoing planning and yet still is timely enough to affect subsequent operations.

In the following sections of this chapter, we propose several ideas for improving the ISR assessment process through better utilization of existing strategies-to-tasks frameworks, standardized mandatory feedback formats, and better use of limited ISRD resources. In addition, we discuss the utility of automated systems to reduce the ISR assessment workload and the need for joint and Air Force doctrinal reform to enable effective ISR assessment.

Suggestions for Improvements

A Framework for ISR Assessment

As shown in the previous chapter, writings on the importance of, and procedures for, ISR assessment are quite extensive, although often nondoctrinal, informal, and in draft form. Despite this body of work, interviews with USAF intelligence staffers from recent conflicts, including OEF and OIF, have indicated several problems with the current assessment system. Comments are generally focused around the lack of a standardized approach in preconflict training, too much emphasis on bean-counting, and too little feedback from ISR consumers during an operation.

To begin, we believe that ISR assessment can best be organized into three unambiguous, logical steps with clear responsibilities at each step. Figure 3.3 highlights the questions to be addressed at each of these three steps. Although the process shown here is simple and may seem obvious, it is critical to separate each of the steps to prevent overburdening any particular portion of the assessment process and setting off in pursuit of potential solutions before the extent of the problem is known. Since few staff are available for a dedicated ISR assessment process, resources must be focused where they are needed the most and needless work avoided.

The first step is the core of assessment and ultimately addresses whether the ISR system is helping the commander accomplish his or her objectives. It does not attempt to capture the "why" if shortcomings are revealed, nor does it include potential prescriptions for improve-

[8] U.S. Central Command Air Forces, 2004; author interview with Dean Daigle, 2005; author interview with Lt. Col. "Gracie" Matthews and Col. Ron Chilcote (ret.), 2005; *ISR Assessment CONOPS Writing Meeting*, 2005; and Poss, 2004.

Figure 3.3
Top-Level Steps in ISR Assessment Process

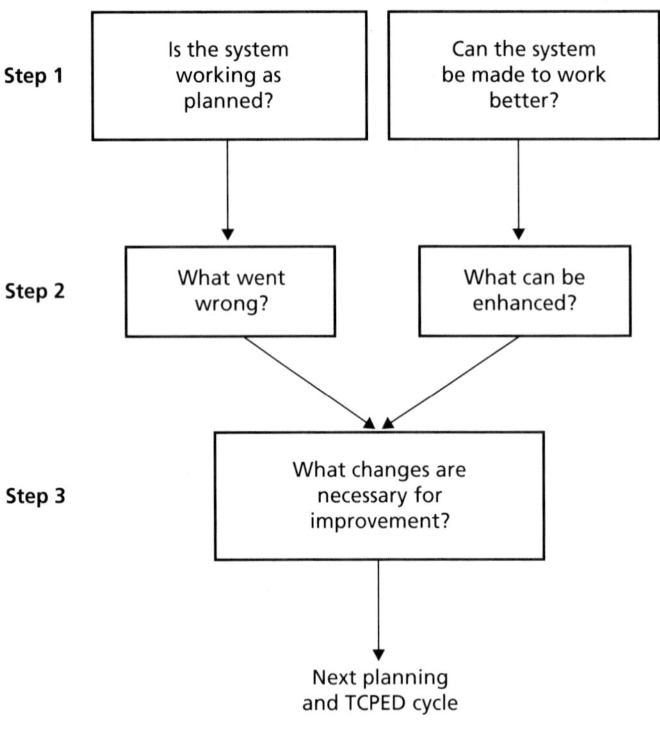

RAND *TR459-3.3*

ment. It is instead a holistic look at the ISR enterprise at the strategic, operational, and tactical levels. This step should be performed regularly, most likely after each ATO cycle at the tactical and perhaps operational levels, and it would involve the system operators, ISRD, ISR personnel in the Strategy Division, and the component or JFC J-2.[9] Much of this step can simply involve data and feedback collection for possible future use.

Step 2 obviously occurs only if issues are raised in the initial assessment. This step is where the detailed information gathered in Step 1 is used to determine the point(s) of failure. It is difficult to specify who should be responsible for this step, since it will be dependent on what performance shortcomings are found. Most likely, the ISRD staff will have the majority of the role here as well.

Step 3 is the obvious follow-on to Step 2 because it builds a prescription for improvement. Again, who exactly should have responsibility for this step will be a function of what problems are found. If, for instance, the problem were with the requestor, for poorly written EEIs for example, then collection-management personnel would be involved to better educate the users. If the problem involved missing or delayed reports, then the PED Cell would likely have primary responsibility.

Since Step 1 is the beginning of, and support for, the entire process, it is worthwhile to define it in more detail. Since many of the questions ISR is trying to answer can be addressed only by fusing information from multiple sources and disciplines, we believe that overall ISR

[9] We use the term "J-2" loosely here, to denote any chief intelligence officer. It would more accurately be described as the S-2, A-2, or G-2, depending on the organization and level.

performance cannot be properly assessed by platform or by intelligence collections. The framework in Table 3.1 attempts to include both the various steps in the ISR process as well as the three levels of assessment highlighted in the ACC assessment functional concept (U.S. Air Force, Air Combat Command, 2005b), although our definitions are somewhat different here. Table 3.1 also highlights the scope of the assessment problem—there are 12 possible assessment areas shown, each with numerous data items and feedbacks required to complete them. Coupled with the requirement to complete at least some operational and tactical assessments within each day's planning cycle, the challenge is evident.

In Table 3.1 we have highlighted what we believe to be the core questions that must be addressed by a useful ISR assessment process. There obviously could be many more; Chapter Two provided other examples from current documents, but those in Table 3.1 should be the

Table 3.1
The Key Questions of ISR Assessment for Step 1

	Strategic	Operational	Tactical
Planning	Are ISR tasks sufficient to support campaign objectives? Are other, nonrequested ISR tasks necessary to support objectives?	Are selected EEIs sufficient to accomplish ISR tasks? Are other EEIs necessary to accomplish ISR tasks?	Are selected collection targets sufficient to satisfy EEIs? Are other collection targets necessary to satisfy EEIs?
Tasking	Are sufficient numbers and types of collection assets available to accomplish ISR tasks?	Are collection platforms being effectively utilized? Are collections being missed that undertasked assets can satisfy? Are nontasked collections preventing satisfaction of EEIs?	What is the fraction of • sensor capacity tasked? • collection platform capacity tasked? • potential collections tasked?
Execution	What is the fraction of ISR tasks completed? Are nonsatisfied ISR tasks preventing accomplishment of objectives? Is accomplishment of objectives being aided or hindered by ad hoc collections?	What is the fraction of EEIs satisfied? Are nonsatisfied EEIs preventing accomplishment of ISR tasks? Are ad hoc collections substituting for the appropriate planned collections?	What is the fraction of planned collections completed? What is the fraction of ad hoc collection slots utilized? What are the timelines of ad hoc collections? What number of preplanned collections are being replaced with ad hoc requests? Are unaccomplished collections preventing satisfaction of EEIs?
PED	Are delayed or missing products hindering accomplishment of ISR tasks? Are delayed or missing products caused by insufficient PED capacity?	Are delayed or missing products hindering satisfaction of EEIs?	What is the • fraction of collections with expected observables present? • fraction of PED capacity utilized? • fraction of collections pushed to requestors? • fraction of collections pulled by requestors? • average time between request for and receipt of a report?

minimum required. Note that the higher-level questions have essentially yes or no answers, and, as discussed earlier, there are two main themes in the table: Is the system working correctly, and can we make it work better? We are not looking to explore potential solutions at this point. As with the ACC functional concept, the tactical-level assessments are focused mainly on data collection with little subjective analysis. Operational ISR assessment should be answering questions using the tactical data, and it is oriented toward platform- and EEI-level performance. Assessment at the strategic level must be focused on whether the component and JFC tasks are getting accomplished and whether those tasks support the campaign objectives. Since shortcomings can occur at any of the three levels, potential solutions can exist in all three as well. We next discuss each in turn, providing more detail on all three steps in the assessment process.

Strategic. Since the ultimate purpose of intelligence is to support the accomplishment of JTF or JFACC campaign-level objectives, it would seem natural to ask how well it is doing that job. Joint Pub 2-0 has a very clear definition of what we consider strategic ISR assessment to be. Unfortunately, strategic ISR assessment is both the most important to conduct and the most difficult to accomplish. During planning, the key questions revolve around matching ISR tasks to objectives. Note that we focus here on ISR tasks rather than PIRs. As discussed later in this chapter, we feel that an explicit strategies-to-tasks framework that specifies the ISR tasks required to support the complete set of strategic and operational objectives is the most appropriate way to guide the collection-management process. Often the JFACC will disseminate such a framework in the daily AOD. We suggest that a similar framework could be used at the JTF level as well, potentially replacing the use of PIRs altogether. As part of this strategies-to-tasks framework, each task would be given a measure of effectiveness to evaluate success. For a task such as "destroy all SAMs," the MOE is fairly clear, but for more difficult ISR tasks, such as "determine the will to fight of indigenous forces," writing measurable MOEs requires a bit more thought. However, the mere existence of written, traceable links between campaign objectives and ISR tasks and the presence of a way to determine whether the ISR task has been accomplished would be a huge step forward. We strongly recommend standardizing this methodology for the AOC and suggest it be employed at the JTF level as well.

Theoretically, if ISR tasks are integrated with objectives and if MOEs are specified for each ISR task, it should be relatively straightforward to assess whether a task has been accomplished and, hence, whether the objectives are being supported. However, much depends on how well the strategies-to-tasks framework is written and whether the MOEs can actually be measured. The framework obviously needs to be complete, in the sense that all necessary ISR tasks to support operational and strategic objectives must be present. Furthermore, the ISR tasks must be measurable, otherwise providing an MOE will not be helpful. For our earlier example of "determine the will to fight of indigenous forces," it could either be accompanied by imaginative MOEs involving communications intelligence (COMINT), or even survey questionnaires, or be rewritten more explicitly so that the MOEs are obvious—perhaps, "determine order of battle of potential indigenous forces." Well thought-out EEIs can obviously play a role here also; our original ISR task could be supported by an EEI counting the number of indigenous force desertions, for example. The main point is that the ISR tasks must be written so that they can actually be accomplished and so that their accomplishment can be measured, preferably in some quantitative manner. Standardized training is, of course, the key to building a cadre of skilled ISR task writers.

In addition to the organization and detail offered by this methodology, it can also provide guidance on weight of effort. The JFACC generally specifies weight of effort for particular ISR tasks, and the PIRs are rank ordered. Although helpful, neither of these techniques addresses the problem of relative importance. In some situations, priority 1 may be 10 percent more important than priority 2, but in others it might be ten times more important. A simple rank ordering or high, medium, low weight of effort scheme does not provide this information. Obviously the JFC or JFACC will make his or her relative weightings clear to the J-2, but often this information does not successfully filter down to the collection managers, Strategy Division, ACF, PED Cell, and SIDO. We advocate explicitly attaching and disseminating a weight or utility to each strategic objective, operational objective, and ISR task in the framework. Not only would this be useful in providing detailed collection-management and dynamic retasking guidance, but these weights can also be used to guide the effort put into ISR assessment. The more important tasks should be examined more frequently and in more detail than the less important ones.[10]

Although the OAT in the Strategy Division is likely the most appropriate venue to conduct strategic ISR assessment, the importance of regular, explicit feedback from the commander cannot be overstated. Since this feedback will probably be given directly to his or her J-2, regular lines of communication between the OAT and the J-2 should be established prior to the conflict and practiced regularly. Venues such as the JCMB could enable communication of feedback from the commander via the J-2, but, in our experience, the OAT does not regularly attend the JCMB, and little time at the meeting is devoted to ISR assessment. The feedback that is provided from the J-2 to the staff typically takes the form, "The boss wants to know why we haven't found those missile launchers yet." While this certainly gets the point across, it is not nearly as useful as, say, a list of the current campaign objectives annotated with how well the JFACC or JFC views that ISR support is being provided to each. It is not hard to imagine a Web-based application containing an up-to-date strategies-to-tasks framework with progress on each ISR task graded by the commander.[11] A complete version could include the EEIs and observables associated with each ISR task. Such tools have already appeared in the Central Command (CENTCOM), for example, based around a PIR-EEI-observable framework, as shown notionally in Table 3.2. As seen, analysts assess the amount of reporting from each platform (or each sensor or type of intelligence collection) toward each observable (on a scale from 1 to 4), and the tool rolls up these "scores" as averages or maxima and applies a color code (black for little or no progress, dark gray for some, and light gray to indicate completion) to give an indication of progress being made toward EEIs and PIRs, as well as to provide a guide toward what platforms are contributing. Obviously, such a tool is only as good as the analysis that forms the judgment and the feedback that reaches the assessor. Whatever format

[10] This approach assumes that all events are independent, though in some instances, this is clearly not the case. We recognize this limitation. A linear, independent approach was chosen for simplicity and transparency, recognizing that it could lead to some anomalous results. A validation task, therefore, needs to be performed, which checks the values assigned to individual targets, deciding whether, as the target level, the relative scores are reasonable. When they are not, experienced analysts may adjust them accordingly.

[11] The ISR Division should already be analyzing progress toward accomplishing EEIs and PIRs and presenting this information at the JCMB. However, our recommendation here is to provide the commander's view of the progress, not the ISR Division's view. If two out of ten EEIs have been satisfied toward a PIR, progress may appear to be poor. However, the commander may have a very different view if, for example, he or she already knows enough to move ahead with other operations.

Table 3.2
Notional Example of Assessment Tool Used by CENTCOM

			Platform 1	Platform 2	Platform 3	Total Score
PIR 1			1.8	1.8	3.0	3.3
	EEI 1.1		2.5	2.5	3.5	4.0
		Observable 1.1.1	1	3	4	4.0
		Observable 1.1.2	4	2	3	4.0
	EEI 1.2		1.0	1.0	2.5	2.5
		Observable 1.2.1	1	1	2	2.0
		Observable 1.2.2	N/A	N/A	3	3.0
PIR 2			3.0	3.5	3.5	3.8
	EEI 2.1		2.5	3.5	3.5	4.0
		Observable 2.1.1	1	4	3	4.0
		Observable 2.1.2	4	3	4	4.0
	EEI 2.2		3.5	3.5	N/A	3.5
		Observable 2.2.1	3	3	N/A	3.0
		Observable 2.2.2	4	4	N/A	4.0

SOURCE: Personal discussion with Ken Ayers, PACAF/A2 Office, July 2005.
NOTES: This framework is notional. Black equals little or no progress, dark gray some, and light gray completion. N/A is "not applicable."
RAND *TR459-Table 3.2*

such tools take, the challenge will be to standardize and make them a core part of the intelligence staff's toolkit through training and regular use.

Strategic-level feedback from other divisions in the AOC or from other service components can also be critical. Many of the campaign objectives that the ISR system could be supporting will involve strike, air defense, or mobility operations within the Air Force as well as maritime or ground operations by the Navy, Army, or Marine Corps. In addition to the tactical-level feedback that these users should be providing about the success of their individual collection requirements, their commanders should also be providing "big picture" feedback on the ISR support necessary to achieve their objectives. With the JFACC operating as the supporting, instead of supported, commander in some of these operations, the feedback mechanisms usually present within the AOC or within the Air Force may not exist. Typically, feedback across these types of divides is accomplished through interpersonal relationships established in previous operations or exercises. Although this informal system can work quite well if all the pieces are in place, it tends to lead to a set of "haves" who know whom to talk with about their problem and "have nots" who do not know how to participate. Potential solutions are discussed below in a dedicated section on feedback, but the key is to formalize (i.e., develop, build service and joint consensus for, document, and train to) a process through which high-level comments, both good and bad, on ISR support toward objectives can be fed back to the Strategy Division and OAT for further investigation.

If problems are found with accomplishing any ISR task in a timely manner or with achieving objectives because of poor ISR support, the work begins on tracking down the source of the problem. At the strategic level, most of the questions should revolve around missing or misstated tasks, weight of effort being put into each task, or perhaps with the force mix available to conduct ISR. The Strategy Division and the various intelligence staffs should revisit the ISR tasks and priorities in the strategies-to-tasks framework to ensure that the list is sufficient to support the objectives, that the tasks are achievable, and that the most important objectives are being supported with high weights of effort. If the framework and guidance appear appropriate, the search must be narrowed down to an examination of the performance against the EEIs under the problematic task. At this point, it will fall to the operational level of assessment to determine where the problem lies and how to solve it.

Operational. Operational ISR assessment is where the "rubber meets the road." Even though shortfalls are noticed at the strategic or tactical levels, often the actual problem and solution will be found at the operational level. If strategic-level objectives are not being accomplished, operational assessment must use the data gathered at the tactical level to identify why and propose solutions. Whether the ISR task is generating situational awareness for the commander or identifying a target for an imminent strike, the most likely shortfalls and solutions are going to lie at the operational level. Solutions are obviously going to depend on the particular shortfall, but usable EEIs, appropriate prioritization of collections, and sensor tasking and synchronization are all likely starting points.

Difficulties at the planning level should most likely focus on the list of EEIs supporting the problematic ISR task. The obvious first step is to revisit, with subject-matter experts, whether the list of EEIs is necessary, sufficient, and relevant to supporting the task. Too many EEIs can be as much an obstacle as too few, since collection capacity will be diluted. Writing usable, relevant EEIs is more art than science, but the same guidelines should apply as those used for ISR tasks. They must be achievable and their accomplishment must be measurable, preferably by quantitative means. If collection managers indicate that poorly written EEIs are inhibiting efficient collection, "feed forward" to the requestors is required to determine exactly what effects are being requested and to educate the requestors on better EEI formats.[12] Each combatant command and Air Combat Command has published guidelines for writing effective EEIs that include examples. Other services likely have guidelines as well. We would recommend that all of these guidelines be reviewed by the ACC/A2 or perhaps the Joint Functional Component Command for Intelligence, Surveillance, and Reconnaissance and a best practices manual be generated for common use and made widely available online. In the best case, such a manual would be approved for joint use, and all potential ISR requestors and managers would be working with the same set of standards.

At the operational-level portion of the tasking process, the importance of correctly prioritizing collections comes to the fore. As discussed earlier, we advocate moving to a strategies-to-tasks utility format to quantitatively link ISR tasks to operational and campaign objectives.[13] With each ISR task given a utility based on top-level objectives, it is straightforward to make transparent prioritization decisions. The same methodology could be carried down to the EEI

---

[12] We discuss later the importance of establishing formal feedback channels from ISR requestors and commanders into the ISR and Strategy Divisions. These feed-forward channels might be equally important, although probably more informal, and it will be the responsibility of intelligence staff to establish necessary relationships prior to any operation.

[13] In this framework, PIRs are relegated to an advisory role or eliminated altogether.

and observable level, with EEIs weighted according to their contribution in accomplishing the ISR task. Such a system could be automated, to some extent, but it is critical to provide the collection managers with insight and decisionmaking authority to make the final prioritization decisions. If EEIs are not being satisfied because they are being considered as low-priority, nontasked collections, that is a sign to revisit the utilities—the values (ratings) that you have assigned to the given collection in the strategies-to-tasks framework. The advantage of this system is that trade-offs in collection priority between EEIs and tasks are clear to everyone and can be intelligently debated so an informed decision can be made.

This prioritization process also plays a role during ISR operations, when collections on the preplanned deck may not be performed because of such operational factors as equipment malfunctions, weather, or dynamic retasking. Having all of the potential collections "apples-to-apples" comparable allows the ISR Operations staff to make appropriate decisions about what targets should be substituted. If these dynamic decisions are preventing (or enhancing) accomplishment of objectives, that information should flow down from the strategic level so that the operational assessment can evaluate the relative utility of the various tasks.

Finally, operational assessment of the PED process should focus on whether processing, exploitation, or dissemination is limiting ISR task or EEI accomplishment. Most of this assessment will rely on feedback from requestors, although some analysis of bottlenecks and capacity shortfalls (communications bandwidth, for example, or number of linguists) can be performed at the tactical level as well.

Tactical. As described in the ACC functional concept (U.S. Air Force, Air Combat Command, 2005b, p. 4), tactical ISR assessment "determines if specific ISR missions provided desired intelligence based on assigned tactical tasks. Key question to answer: Were the individual tasks assigned to ISR operators/assets performed successfully?" The focus on specific missions clearly limits this assessment to the sensor and perhaps platform level. It is not clear what is meant by "tactical tasks," but we assume this implies executing the assigned collection deck or disseminating a sortie's intelligence reports, for instance. We agree with the U.S. Air Force, Air Combat Command, 2005b: Tactical assessment should be focused on collecting quantitative data rather than on making subjective judgments. The main focus here should be on the performance of specific systems—whether they are doing all they are asked and whether they could be performing better. Feedback from system operators is particularly critical in answering this latter question since they will probably be much more knowledgeable about the full capabilities of their system than the AOC will be. Postmission reporting can provide some of this information as well, although typically much manual effort is required to extract useful information and populate databases with it.

The main exception to the quantitative nature of tactical assessment occurs when assessing the planning portion of the intelligence cycle. At the lowest level, collection planning relies on subject-matter experts (typically from the ACF) and preconflict IPB to turn EEIs into observables and finally into specific signals to collect or locations to image. This is a very "tactical" task, but the ability of ISR to satisfy the commander ultimately relies on performing this task well. Unfortunately, it is also difficult to assess. Given the difficulties in attempting to determine a priori whether the observables to be collected are sufficient to satisfy every EEI, it is probably more efficient to utilize the assessment of tactical planning in a top-down manner from the operational and strategic assessments. If sufficient progress is not being made on a particular EEI, then the problematic EEI can be revisited and the choice of observables and collections examined in more detail, perhaps with a broader range of subject-matter experts from other locations.

It is important to remember that perfect tactical ISR assessment may not provide many useful insights at the operational and strategic levels. As already discussed, tasking, collecting, and disseminating 99.9 percent of what was requested is not very helpful to the commander if the wrong targets were collected or reports arrived too late to be useful. One good, high-priority collection is worth more than 100 poor or low-priority collections, and it is up to the operations and strategic assessment efforts to determine whether the collections have high utility overall. However, if problems are found during higher-level assessments, the data collected here are critically necessary in determining what is not working properly and how to fix it (Steps 2 and 3 of the assessment process).

Since tactical assessment functions mainly in a supporting role, it is important to avoid devoting too many resources to it. We would advise making full use of statistical sampling and automatic data collection and archiving to collect the necessary information. The personnel in the flying squadron and ISRD who are most likely performing this task are too useful in other roles (including operational ISR assessment) to spend much of their time laboriously filling in spreadsheets. A fairly simple Web-based form linked to a database could collect and store all of the necessary data, most of which would not be viewed unless problems were highlighted by higher levels of assessment. Some simple automated analysis could be performed on the raw data to initially flag unexpected results or negative trends for follow-on analysis.[14] The idea is simply to monitor that the collection systems are functioning correctly. The overarching theme, however, should be to minimize the personnel hours spent entering and processing raw numbers.

Feedback. As discussed earlier, any framework for ISR assessment must address the fact that many of the questions ISR is trying to answer are subjective and qualitative in nature. PIRs about an enemy's intent, for example, can be objectively answered only by interviewing the participants, presumably after the operation is over. Similarly, PIRs that require every SAM or ballistic missile launcher to be located can never truly be assessed as accomplished—how do you know whether one was missed? Additionally, simply counting collections, even in increasingly complicated ways, is no measure of ISR success. Quality is a better measure of performance than quantity. As a result of this inherent inability to quantitatively measure success, the ultimate arbiter of ISR success must be the intelligence consumer. At the highest level, only the joint forces commander can judge whether the intelligence provided has supported the accomplishment of his or her strategic objectives. This requirement places feedback at all levels, from tactical requestor to the ISR division and from the JFC to his or her J-2, as the ultimate source of ISR assessment. The challenge is to gain this feedback in a timely manner in a usable format, correlate and integrate it with quantitative measures, and use it to guide subsequent collection efforts.

We believe there are two critical components to gaining and using feedback on ISR performance. The first is clear and unambiguous instruction from the highest levels that feedback on ISR performance is required of all ISR consumers. Collection managers, intelligence analysts, and personnel in the ISRD do not have the authority to require feedback; only task force and component commanders can do so. Whether implemented in the operational plan, theater

---

[14] The idea here is similar to a warning light in an automobile. Unless problems are detected, no attention is required.

special instructions, the daily AOD or via message, the joint forces commander and JFACC must make it a requirement of ISR requestors that part of their responsibility includes providing feedback. Just as they have a responsibility to follow approved channels and use approved forms and terminology for requesting ISR support, feedback must be included in this responsibility. Similarly, this requirement for commanders and obligation on users must be explicitly reiterated in the joint and service doctrine. Although it already appears in several doctrinal and TTP documents, the message does not appear to be getting through.

The second critical component to enabling better feedback supports the first. The ISRD, in conjunction with the JTF J-2, must provide a standard, theater-wide feedback form and channel. Standard forms and systems are used for collection requests; the same should be done for collection feedback. Currently, for example, the CENTCOM ISRD provides simple Web-based forms for users who wish to provide feedback. This format has advantages, including easy access for most, but not all, users. However, if made compulsory, such a Web form will need to be supported by a robust database management system to handle the much greater quantity of feedback that would be expected. Another possibility is to provide a link to a standardized feedback form within the PRISM collection-management system. Since many users are requesting collections here, they could also provide feedback on the previous cycle's requests at the same time as submitting new ones. Because PRISM is already a Web-based application, this should not be difficult to implement and integrate into the preexisting PRISM database.

Perhaps the most important source of feedback, however, is the component and JTF commanders themselves. Since it would not seem appropriate that they provide Web-based feedback, this feedback will likely be face-to-face from the commander to his or her intelligence chief, the J-2. Since the purpose of ISR is to support the objectives of the commander, and the only direct way to determine whether it is doing so is by asking the commander, this feedback process is the most critical and, yet, the most dependent on a personal relationship. We suggest that the process be formalized to the greatest extent possible, with assessment feedback being a regular, scheduled part of the J-2's briefing to his or her commander. Specific questions (i.e., the strategic questions from Table 3.1) should be regularly asked of the commander by the J-2, and the answers should be formally disseminated to the staff. Much effort is put into disseminating the commander's intent throughout the headquarters staff—similar effort should be put into disseminating the commander's feedback. Since senior-level feedback should also be generated at other service components and intelligence organizations, the JTF J-2 staff should establish formal feedback dissemination channels (regular videoconference, for example) between the primary receivers of that feedback (component J-2s for the services) and the JFACC's J-2 and the ISRD in the AOC, either directly or by acting as a conduit.[15] Since the top-level ISR feedback process will be most important during conflict operations, it needs to be discussed, agreed upon, and practiced by the commander and J-2 beforehand. Making this feedback step a part of J-2 and senior leader exercises and training will ensure that there are no surprises for either the commander or intelligence officer when real operations begin.

We conclude with a summary of the primary observations and recommendations made throughout this chapter:

---

[15] In the end, this high-level feedback must make it back to those who can affect the ISR system. To us, that appears most likely to be the JTF J-2 staff, the JFACC's J-2, and the ISR and Strategy Divisions in the AOC.

- An ISR assessment process is critical for determining how well ISR is supporting campaign objectives.
- Poor performance by the ISR system can affect the conduct of the entire campaign.
- Air Force and, to a lesser extent, joint doctrine provides little or no guidance on performing ISR assessment other than to mention that it should be done. AFOTTP 2-3.2 provides, by far, the most detailed and useful guidance on ISR assessment.[16] This guidance plus recent work by ACC and current combatant command best practices should be utilized in a bottom-up manner to form the next revision of Air Force ISR doctrine. Joint discussions should also be held to compare techniques across services in preparation for joint doctrine revisions.
- Adopting a strategies-to-tasks framework for collection planning at the JTF level will enable useful strategic and operational ISR assessments because ISR tasks will be clearly linked to campaign objectives and accompanied by measures of effectiveness. Because the framework will need to be available to both ISR requestors and operators, the PRISM or Collection Management Mission Applications system could potentially be used to link collection requests to ISR tasks and, hence, to objectives rather than merely to PIRs. Further, if the ISR tasks and objectives are weighted in importance rather than simply ranked, more-intelligent decisions on the appropriate level of effort devoted to collection and assessment could be made.
- Standardized, joint manuals for the writing of measurable ISR tasks, EEIs, and observables should be generated and disseminated by the ACC A2 or the Joint Functional Component Command for Intelligence, Surveillance, and Reconnaissance using best practices from current efforts in this area by the various combatant commands and components. ISR operations personnel at all levels should incorporate these manuals into initial and continuation training programs.
- Database management[17] systems should be used to manage the collection and processing of quantitative ISR performance data. Since some of this information will be provided through feedback, a common database is probably the most efficient mechanism. Statistical sampling, standardized formats, and automatic flagging of unexpected results could be used to reduce the amount of manual labor.
- JTF J-2 staff and/or the ISRD in the AOC should develop and disseminate standard Web-based assessment forms for all requestors and users of ISR-generated intelligence. These forms could also be linked to the intelligence request process, perhaps through the PRISM system. A standard database management system could be used to effectively manage this information and capture the qualitative meaning of the feedback data.
- JTF and component commanders should mandate feedback on ISR performance from all requestors and users of ISR-generated intelligence. Service and joint doctrine as well as training curricula should reflect this requirement.
- Prior to operations, senior members of the JTF and JFACC intelligence staffs should plan to elicit feedback from their respective commanders on the ISR contribution toward achieving objectives.

---

[16] AFOTTP 2-3.2 is being revised and is in the coordination process. A draft dated September 2006 will be published as AFOTTP 3-3.6. We refer to the most recent approved document.

[17] Perhaps more broadly "knowledge management," but a good first step would simply be organized and accessible database systems.

A Framework for Allocating ISR Resources

A strategies-to-tasks framework is well suited for identifying the complete range of operations that could help satisfy the commander's PIRs. Furthermore, the value could be extended to include links between "finders" and "shooters" through concepts of execution (CONEX) for accomplishing almost any operational task. This framework should also help enable effective decentralized execution based on the guidance given from senior leadership (centralized control). The framework consists of campaign-level objectives, operational objectives, and operational tasks. As noted in Chapter Two, there is already a strategies-to-tasks framework utilized in the AOC, namely in the Strategy and Combat Plans Divisions, as part of the ATO production process. The teams that make up these divisions use this framework to create target nominations that achieve the commander's objectives. Note the similarity with the ISR planning process. Both divisions are taking top-level commander's guidance and forming a list of targets, although in the ISR case it is a list of collections. Rather than the two processes using two different sets of objectives and tasks, we suggest they coordinate their efforts and use a single, unified framework, informed and expanded upon by the commander's PIRs for use in ISR allocations in support of the overall campaign planning.

The top level of this framework is the commander's strategic, theater-level objectives—those that are essential to achieving positive campaign outcomes. Examples of these objectives include such top-level statements as "halt the invasion" and "protect U.S. and coalition troops." Under each of these strategic objectives is a set of operational objectives to be achieved to help support the top-level campaign objective. These operational objectives will probably need to be expanded upon from the targeting framework to support all of the ISR requirements. For example, "gain air superiority" and "monitor weapons of mass destruction (WMD) activity" might be two examples of operational objectives that fall under the campaign objectives described above. This second objective is not one that would be expected to appear in a targeting framework. Instead, it would be added to the framework as the result of a commander's PIR, such as "will the enemy employ WMD?" See Figure 3.4. Furthermore, for these ISR-specific strategic or operational objectives, the EEIs associated with each PIR can easily serve as a guide to appropriate operational objectives or operational tasks. We have noted what might be additional objectives and tasks in Figure 3.4 with italicized print and thicker borders. The change from current processes here is that PIRs and EEIs guide modifications to a preexisting strategies-to-tasks framework rather than form the top level of an ISR-unique framework.

Using PIRs to Guide Development of ISR-Specific Objectives and Tasks

Operational tasks are at the lowest level and could include such tasks as "observe suspected storage sites" to support the WMD-related operational objective described above. In Figure 3.4, we highlight that below each of these operational tasks are the CONEX necessary to actually accomplish the task. The CONEX use the observables of each operational task, such as, "use GMTIs [ground moving target indicators] to monitor traffic to and from suspected chemical weapons sites" and "take EO [electro-optical] imagery to identify activity consistent with movement of chemical weapons," to guide the specific collections to be performed.[18]

[18] Subject-matter expertise on the enemy's behavior and various sensors contributions to intelligence are necessary to perform this job, particularly for difficult targets such as WMD.

Figure 3.4
Commander's PIRs and Development of ISR-Specific Objectives and Tasks

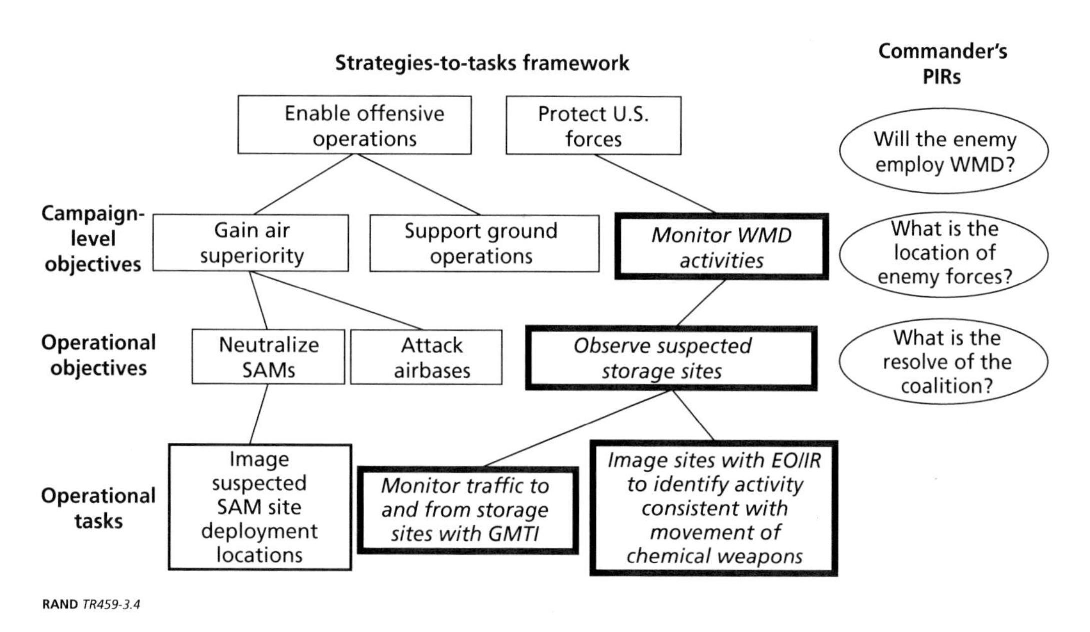

RAND *TR459-3.4*

Note that in Figure 3.4, we associated a particular sensor type (GMTI and EO/infrared (IR)) in our exemplary CONEX. Doing so could be optional, but it does bring advantages. First, some types of sensors may provide more information than others, and so they would be preferred. By pointing them out, more capable sensor types can be given higher priority. Second, this methodology easily allows for multiple sensors to be focused on the same target, which may improve probability of collection and enable advanced processing techniques. The disadvantage of this approach is simply the additional workload of generating utility values and managing the large number of operational tasks that may result.

A strategies-to-tasks framework for ISR could be useful for a number of reasons. To begin, it helps to identify a range of collection strategies for satisfying the commander's objectives. Using this framework will help to identify a range of effects-based options. Such a hierarchical list also makes it possible to trace those tasks at the lowest level back to the commander's intent. Lt. Col. Daniel Johnson (2004) identifies both of these advantages in his paper. An additional advantage of the framework depicted in Figure 3.4 is that the collection requests driven by targeting and intelligence needs are both present. Rather than requests for ISR support being passed "over the transom" from the Combat Plans Division, the ISRD now knows exactly what the planners are trying to accomplish and has a transparent audit trail for how they support top-level objectives. In addition, the different communities planning future offensive operations and future ISR operations will be able to better "speak the same language," allowing better integration across the AOC. The final advantage of this hierarchical framework is that it simplifies the process of assigning priorities to various collection tasks if a utility metric is used at each stage of the process.

Assigning Utility to Objectives and Tasks

Current doctrine speaks of assigning priorities to requirements. The problem with a simple ranking or prioritized list is that it does not allow identification of the relative worth of a higher-priority collection when compared with one or more lower-priority collections. Furthermore, because no utility function has been assigned to the objectives, it is difficult during the execution phase to weigh various potential ad hoc taskings against planned tasks. For example, should two low-priority, time-sensitive collections be substituted for a single, preplanned, higher-priority collection? By assigning utility values or relative weights to all potential collections, better guidance can be provided for those making decisions about retasking sensors or assets.

At this point in the development of our utility framework, we should consider the utility of a successful collection only when assigning utility values. Later, we will take into consideration whether a collection can be made with the available assets and the probability of a successful collection. Both the utility of the collection and the probability of making a *successful* collection will be considered when making decisions about planning a collection strategy and making decisions about retasking assets. However, at this stage, we are concerned only with identifying the utility of various collections.

The process starts with the campaign-level strategic objectives. A set of objectives is defined, most likely at the JFC level, and weights corresponding to the relative importance of these tasks are then assigned. The Strategy Division could certainly play a role in this effort. The sum of weights across all the campaign objectives would be normalized to 1.[19] Note that we are *not* making resource allocations at this stage, but rather identifying the utility of achieving these campaign objectives. Initially, the utility of these objectives will come from information generated by IPB, but it will evolve over time as the campaign progresses through various phases and our understanding of the adversary improves.[20] In fact, preconflict deliberate planning could map out the weights of each objective and task for every campaign phase.

Next, a set of operational objectives that helps to achieve each campaign-level objective is identified. This task is best performed at the command level, advised by subject-matter experts who understand the adversary's behavior. Good IPB is needed to prepare a set of operational objectives to best serve campaign-level objectives. Weights are assigned according to the contribution of each operational objective toward accomplishing campaign-level objectives. Again, the reader is reminded that we are not assigning level of effort at this stage. We are simply identifying the contribution of each successful operational objective to a campaign objective. The weights of each group of supporting operational objectives under a single campaign objective should be normalized to sum to 1. Since the ISRD will be utilizing parts of the framework from the Strategy and Combat Plans Divisions, their guidance on utility values is important, although the ISR planning process may be using different weights at the lower levels than the targeting process uses, since it could have a somewhat different set of tasks (those that are intelligence only, for instance).

Finally, at the lowest level, a set of operational tasks that support each operational objective should be defined. All of the collections that could end up on the JIPCL will eventually be associated with an operational task. As before, each operational task should be assigned a weight that corresponds to the contribution of a successful task toward its corresponding operational objective. All the tasks under an objective should sum to 1. The weights assigned at each level of the hierarchy should be evaluated on a regular basis (i.e., every ATO cycle), along

[19] Normalizing to 1 allows comparison across the campaign objectives and subsequent operational objectives and tasks.

[20] In some instances, our understanding of the adversary's intent could worsen, for example, if the adversary's intentions change.

with consideration of any new objectives or tasks. The JCMB could be a good forum for this discussion. If the CONEX under the tasks contain several approaches that could compete for resources, weights can be applied here as well.

By multiplying out the weights through the hierarchy for each of the tasks (see Figure 3.5), the *total utility* of successfully accomplishing a task can be quickly identified. In this example, the collection utility of 0.112 under the "Image suspected SAM site . . ." task is obtained by multiplying the strategic objective utility of 0.4 ("Enable offensive operations"), the 0.8 utility of the operational objective ("Gain air superiority"), the 0.7 utility of the operational task ("Neutralize SAMs"), and finally by dividing by 2 for the two collection targets that support the task. The total utility is a campaign-level measure of the relative utility of individual tasks. Likewise, if all of the collections associated with a task are assigned a utility summing to 1, the total utility of each individual collection can be found by multiplying by the task utility. The additional workload of assigning weights to all of the objectives and tasks should not be onerous, given that all of them are already placed in rank order in the current process. Additional thought will certainly be necessary to decide how much more important higher-ranking objectives are than lower ones, but good planners already consider these factors. The advantage here is that this thought process will be formalized and made transparent to high-level commanders, their ISR planners, and others who request ISR support.

Figure 3.5
Calculation of Notional Target Utility Values

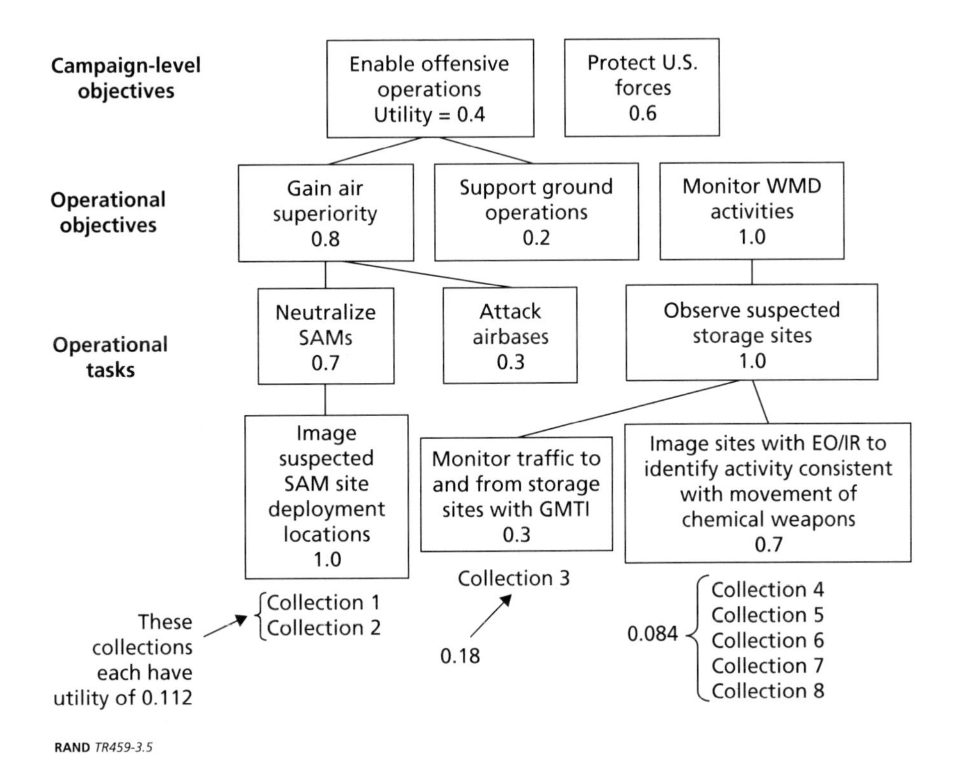

RAND *TR459-3.5*

Notional Target Utility Values

Note that objectives with many tasks or tasks with many collections could result in lower total utilities for each collection, since the sum at each level must total 1. In the example in Figure 3.5, while Collection 3 is associated with a lower-priority element of the CONEX than Collection 4, the fact that the entire task can be satisfied by that single collection is reflected in the higher utility value associated with Collection 3. Also, note that there may very well be duplicate collection targets that satisfy more than one operational task. In this case, the utilities for each occurrence can simply be added together to produce a higher utility. This process emphasizes the efficiency of collecting on targets that help achieve multiple objectives. When all of the utilities have been calculated, the result should be a prioritized list of targets.

This formulation for computing utility has several benefits. First, there is a clear and direct link for the value of each individual operational task and collection at the lowest level all the way up to strategic objectives. When tasking and retasking decisions are made, it is possible to quickly calculate the comparative values of various collections. Second, the hierarchical nature of the process makes for a natural division of labor across the chain of command. Senior leadership can remain focused on the relative importance of top-level strategic objectives and ensure that the utility values are correct, while specialists and subject-matter experts can work on operational tasks and the collection targets that will support them. Third, the process is able to quickly accommodate changes in a commander's guidance or unexpected adversary behaviors. When changes are made to the utility weights at the campaign level, recomputing the total utility for each of the operational tasks can be done very quickly, leading to a rapid reprioritization of the collection list. In addition, changes to collections to improve the accomplishments of operational tasks or objectives during the campaign can be easily performed. This factor is important because it preserves the weight of military judgment for forming a prioritized list of targets.

One other, less quantifiable advantage of this framework is that it helps to emphasize the operational level of ISR strategy, planning, and collection management. By allowing the intelligence staff to plan against the same objectives and tasks used by the strike planners, the two components in the AOC should more easily integrate and begin to see new opportunities for supporting each other.

Probability of Successful Collection. The importance of an individual collection is the primary factor considered by collection managers when planning ISR operations. However, collection managers also have other potential factors to consider when planning operations, such as

- poor weather between an EO/IR sensor and the target
- low grazing angle
- target relocation
- terrain obscuration
- effective concealment, camouflage, and deception
- short duration, rare signal emissions
- encrypted signals.

These individual considerations fall into the category of factors that affect the probability of achieving a successful collection against an individual requirement.

In this subsection, we propose a method for incorporating the probability of successful collection into the strategies-to-tasks utility framework. Despite the difficulties in determining extremely accurate probabilities for each collection, including estimates of the effect of such factors may aid in the creation of a more realistic collection strategy. Without accounting for such effects, the prioritized collection deck created by the previous method will be more of a collection "wish list" rather than an operationally relevant list.

After utilities have been assigned to each operational task, the next step in the planning process is to assign the available collection assets to maximize the expected collection utility for that day. Using the process described in the previous section, we assign utilities to each operational task. Then, we estimate the probability of successful collection based on the system capabilities and a thorough knowledge of how the adversary behaves.

While the probability of successfully collecting against any single target cannot be known exactly, it should be able to be estimated. In addition, while the probability of successfully satisfying a single request may be low based on a single collection, that probability might be increased by making multiple collections with a single asset, by persisting over a given target area, or by performing collections with multiple sensors or platforms.

The expected utility of any given collection is defined by the product of the total utility of the probability of success with the probability of success given that collection. Mathematically, this is

$$E(j) = U(j) \times P(j),$$

where $E(j)$ is the expected utility of collection j, $U(j)$ is the calculated utility of collection j, and $P(j)$ is the probability of successfully achieving collection j.

During the deliberate planning process, collection managers should strive to form a collection plan that maximizes the expected total utility for that day. In other words, the collection managers should strive to maximize

$$\sum_i U(i) \times P(i),$$

where the set of collections i is limited by the number of assets, types of sensors, and time. While this may seem to be a difficult optimization, remember that it is likely that a certain number of our collections may simply be infeasible (because of poor weather or long range, for instance) and, therefore, have zero probability of success. Those collections should be removed from consideration prior to the optimization process. Furthermore, it is not necessary that an exact probability of collection be assigned to every target. Simple categories such as high, medium, and low, with associated numerical probabilities such as 0.9, 0.5, and 0.1, might suffice for many collections. If the operational tasks are defined as a function of sensor type, the tools implementing this construct could easily be created so that targets satisfying multiple operational tasks with multiple sensors get a boost in their probability of collection.

The Effect of the Methodology on Deliberate Planning Processes. The ultimate output of the strategies-to-tasks framework laid out here is the daily collection deck for each ISR platform. As just mentioned, the objective is to maximize the expected utility of all the collections. There are many options for forming a collection strategy with this methodology. With the utilities and probabilities of collection in place, a prioritized list of collection targets can be calculated. Planners could simply start at the top, with the highest expected utility collections, and work their way down until all of the collection capacity has been tasked.[21] Targets with high priority but low probability of collection (ballistic missile launchers, for instance) will not consume excessive collection resources. The disadvantage to this approach is that in a capacity-limited case, it is possible that targets supporting some operational tasks might never be serviced, which could also be a problem with targets having a very low probability of collection. If situational awareness, rather than collection of information against specific targets, is desired, then a task associated with gaining situational awareness about a particular area could be defined and assigned the appropriate utility in the framework.

Another method would be to skip the final step of calculating the utility of every target and simply calculate the utility of all of the operational tasks. Since the sum of these utilities must add to 1, the utility can be converted directly into a percentage of collection capacity. For the example in Figure 3.5, the "Image suspected SAM . . ." task would get 22.4 percent of the collection deck, the "Monitor traffic . . ." task would get 18 percent, and the "Image sites . . ." task would get 42 percent.[22] This method has the advantage that every task will be collected against, but the disadvantage that targets associated with higher-priority tasks may be rejected in favor of lower ones. With our example, if the number of targets associated with "Monitor traffic" requires more than 18 percent of the capacity, should these targets be rejected in favor of some for the "Image site" task? Such decisions are probably best left to the judgment of the planners. The advantage of pursuing a utility methodology is that all the information is available for intelligent decisionmaking. The most likely tool for implementing this framework in the deliberate planning process is PRISM (the Web-based tool discussed above). It is currently used to integrate collection requirements from the JFC and various components and, with other tools, to generate the daily collection deck. It would be a reasonable modification to add the strategies-to-tasks and utility functionality discussed here. Each collection target would be attached to higher-level operational tasks and objectives instead of to PIRs. The utility functions attached to each of these would be updated every day and used to calculate the task and collection target utilities. The desired methodologies for building the collection deck could then be implemented as algorithms within PRISM, available for use by the collection planners.

The Effect of the Methodology on Dynamic Retasking Processes. During execution of the collection plan and the day's ATO, sensor assets are routinely retasked in order to collect against ad hoc requests and other unplanned collection opportunities that may present themselves. Guidance for these retaskings comes from the RSTA annex of the ATO, which lists the commander's priorities for collection each day and often specifies in which situations assets can be retasked. An example of such a collection would be finding the location of a downed pilot to support a search-and-rescue operation. Since most of the capacity of collection platforms is filled during the deliberate planning process, changing the plan for an ad hoc collection may require losing other collections. The added step of moving a collection platform could result in losing even more planned targets. Retasking decisions are made on a regular basis by the staff of the ISRD on the AOC floor, operators in the Distributed Common Ground Station, and occasionally by sensor operators themselves. All are doing their best to interpret the command-

[21] Once all of the collection requirements have been ranked, an initial step would be to separate out those to be collected by different sensors and platforms, since they do not compete with each other for collection capacity.

[22] These do not add up to 1 because we have omitted some operational objectives and tasks from our example.

er's intent but have little quantitative information available to them to determine whether new collections are more important than planned ones.

With this strategies-to-tasks utility framework, when making decisions about retasking assets, we can use a relatively simple mathematical test. If the expected utility of the new ad hoc tasking is higher than the expected utility of the original tasking(s), then that retasking should be performed. Mathematically, this means retasking should be performed if

$$\sum_i U(i) \times P(i) > \sum_j U(j) \times P(j),$$

where the i tasks are associated with the ad hoc tasking and the j tasks are associated with the originally planned tasking. Ad hoc taskings should receive priority only if the new collection utility and probability of success give, overall, a better solution to addressing the commander's objectives.

One benefit of this framework is that it makes retasking decisions relatively straightforward. Take a leadership target, for example. Searching a city for this target with EO sensors would have a low probability of success without any other supporting information; therefore, the expected utility, which includes probability of success, associated with that task would be very close to zero. An unmanned aerial vehicle equipped with EO sensors would probably be put to better use performing other tasks on the battlefield, even if the task of finding the leader has a high utility associated with it. However, if we receive a good cue from another source that a leadership target is in a given village, at this point, the calculus associated with retasking changes because of the higher probability associated with the target. Good ISR operators and collection managers already utilize this thinking in their decisions, but our proposed framework adds some formality to the process.

To implement this retasking methodology, the ISR Cell on the AOC floor must have access to tools capable of displaying and recalculating the relative utilities of the planned collection deck and potential ad hoc collections. This requirement highlights the need to be able to rapidly place a previously unknown target into an operational task and to assign a probability of collection. To do this, target categories must be planned for in advance with operational tasks in the framework. An example is the retasking required to accomplish the operational task of rescuing a downed pilot. One of the CONEX for this task would be to "image site of downed aircraft," which has no collection targets assigned to it for deliberate planning purposes but would have a high utility attached so that ad hoc requests could be quickly accomplished. Other emerging targets, such as missile launchers or leadership targets, would probably already have appropriate tasks in the framework.

In addition to the utility functions, the AOC staff also needs tools to visualize the real-time location and sensor capabilities of the available ISR platforms in order to choose the most appropriate system for retasking. This functionality could be a part of the Collection Management Mission Applications, but these tools must be fully integrated into the AOC to provide the needed information rapidly enough. In addition, these tools need to be linked to Strategy Division tools and output used to plan the strategies-to-tasks framework associated with air operations.

CHAPTER FOUR

Collection Requirements Tool

In the previous chapter, we examined comments made by the military on recent operations with regard to ISR planning, tasking, and assessment of ISR asset use. Recommendations were presented to improve assessing how well ISR assets are employed to support the commander's objectives and other combatant command's intelligence collection needs. We also suggested a utility-based strategies-to-tasks framework to align the ISRD collection process with that of the combat plans. We now turn our attention to a quantitative approach for assessing the costs[1] and benefits of alternative collection strategies, followed by a discussion of measures of performance and effectiveness to make a judgment on how well a particular strategy performed.

This study's scope necessitated the development of two analytic tools to examine the tasking and employment of ISR assets. The analytic framework is designed to address the client's primary concern at the time: how to quantitatively express the costs and benefits of using a particular ISR collection strategy rather than another one. Figure 4.1 illustrates how the two tools fit together.

The first tool, shown on the left, focuses on the deliberate planning and scheduling of ISR assets to support combat operations and the commander's need for information. With this tool, we can experiment with how sensors are allocated to meet collection requirements and investigate the assignment of requirements to sensors based on different collection strategies.

The first tool assumes that several parameters are fixed. For example, ISR aircraft orbits are stationary and do not vary during a given ATO day.[2] Also, airborne platforms fly with their current ISR sensors. Finally, the collection deck is limited in capacity for imagery collections based on existing exploitation limitations. The output of the collection requirements tool (CRT) is a series of collection decks for each platform and sensor combination.

The second tool models the employment of ISR forces to meet the requirements of the scheduled collection decks (which come from the CRT) and the need to collect information on pop-up targets or areas of interest, including both time-sensitive and ad hoc targets. This model allows the analyst to examine the costs and benefits of tasking ISR resources using alternative collection strategies. He or she may also explore the effect of retasking assets for ad hoc targets instead of pursuing planned collection deck requests.[3]

---

[1] *Cost* here refers to the loss of collections on other targets for the sake of those targets on which collections are made.

[2] This tool can easily be expanded to account for ISR satellite orbits as well.

[3] The fiscal year 2005 modeling focuses on the collection process. However, the model captures the PED process as well and would prove a useful tool for examining issues associated with exploitation and dissemination.

Figure 4.1
Analytic Framework for Quantitative Assessments

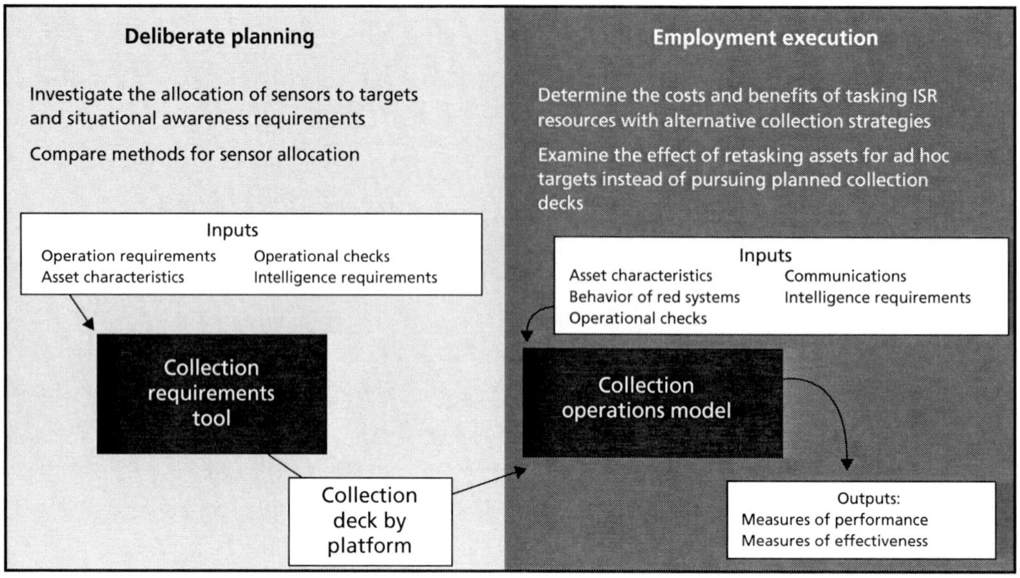

Deliberate planning

Investigate the allocation of sensors to targets and situational awareness requirements

Compare methods for sensor allocation

Inputs
Operation requirements
Operational checks
Asset characteristics
Intelligence requirements

Collection requirements tool

Collection deck by platform

Employment execution

Determine the costs and benefits of tasking ISR resources with alternative collection strategies

Examine the effect of retasking assets for ad hoc targets instead of pursuing planned collection decks

Inputs
Asset characteristics
Communications
Behavior of red systems
Intelligence requirements
Operational checks

Collection operations model

Outputs:
Measures of performance
Measures of effectiveness

RAND *TR459-4.1*

The collection operations model (COM) also assumes that some parameters are fixed. In addition to the assumptions for the first tool, the COM has scripted or fixed red forces behaviors during the model run. Note that it is not the model environment that constrains red forces behaviors; rather, we currently impose on it scripted behaviors. By fixing red forces actions, we can focus our analysis on how blue forces perform in observing red forces based on varying blue forces collection strategy.

The output from the COM is a series of performance and effectiveness measures that we discuss in Chapter Six.

Inputs for the Collection Requirements Tool

In the CRT, assets are assigned to meet a defined collection strategy in four steps as illustrated in Figure 4.2. We briefly outline the steps here and, in subsequent sections, elaborate on particular steps in the process. The first step is to establish the collection requirements. They consist of direct operational support requirements and broader intelligence requirements. The former include such tasks as positively identifying a target before a strike package drops weapons on it or providing combat assessment (e.g., BDA) on a target previously struck. Operational support requirements can help the targeting process both pre- and postmission. Often, these requirements lead to bean counting when assessing ISR's role in the mission.

Intelligence requirements tend to be broader in scope. They address the commander's information needs regarding the enemy's intent, blue force protection, and situational awareness, to name a few. An example would be, what is the enemy's intent for using chemical or biological weapons? For this example, evaluating the quality of intelligence information is difficult until the enemy can be interviewed postconflict.

Figure 4.2
Collection Requirements Tool Process

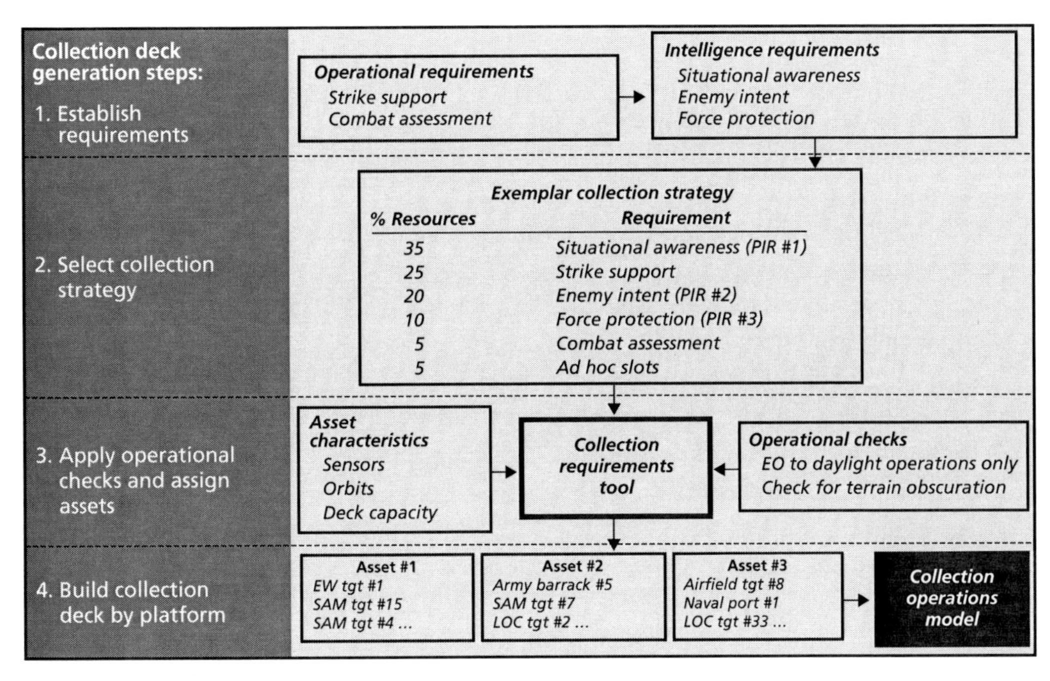

RAND *TR459-4.2*

After the requirements are known, a collection strategy is selected. One example from operational experience is when a portion of ISR assets is assigned to each requirement. Typically, room is left available on collection decks for ad hoc targets that appear after the planning process. In the previous chapter, we presented a collection strategy based on a strategies-to-tasks framework using utility-based weighting that could be used here.

The next step is to apply operational rules to collection assignments for meeting the prescribed requirements. An example would be to select EO sensors for daylight operations only. To assign assets to meet requirements, asset characteristics, such as sensor parameters, the orbits the platforms fly, and the deck capacity associated with each sensor and platform combination, must be accounted for. All data are fed into the CRT, which produces a series of collection decks for each platform and sensor combination. The decks are subsequently fed into the COM, which is discussed in the next chapter. The CRT is best suited for classes of targets that are fixed, are associated with fixed locations, or have known deployment sites. An alternative approach for other classes of targets would need to be employed and is not explored in this study.

The ISR modeling process is centered on the System Effectiveness Analysis Simulation (SEAS) modeling environment. This tool simulates the operations of both sides of a conflict and includes sea, ground, and air units, as well as third-party "green" units. The COM uses SEAS to model the processes involved with ISR, including TCPED. Since SEAS does not include the functionality to model some of these processes, it was necessary to preprocess some of that information to feed into the SEAS model. The following two sections discuss the tools and techniques used to convert a list of potential ISR collection targets—identified by name, type, and location—into a set of collection decks to be used by each sensor on each platform.

Operational Considerations

Blue Force Characteristics. Collection managers and associated personnel who have the responsibility of generating collection decks for blue force ISR assets have to consider numerous factors:

1. The commander's PIRs and the relevant operational and tactical ISR objectives.
2. The blue force ISR capabilities, that is, the virtues and shortcomings of ISR platforms and their associated sensors. Questions that have to be addressed include the following: Is the target within the maximum range of the sensor? Can the sensor provide the required geolocation accuracy? Can the IMINT sensor achieve a target image with enough resolution to detect, classify, and/or identify the target? (For the deliberate planning of collection decks, we assume an average terrain obscuration such that sensors are limited to a 6-degree grazing angle.) Can the electronics intelligence (ELINT) sensor be tuned to the expected frequency of the target's emitter? Can the ELINT sensor analyze the signal characteristics expected for that emitter?
3. Geometrical considerations such as whether the target is within line of sight (LOS) with respect to the ISR platform. The first factor that causes loss of LOS between the platform and the selected target is the curvature of the earth, that is, the target may be located beyond the horizon. The second factor is obstruction due to the local topography around the target area. For example, hills, mountains, and/or tall buildings may surround the target, or it may be located within a basin or geological depression.
4. The characteristics of the target, namely, its geographical location (if known), its size (given either as a volume or a number of elements), whether it is mobile or stationary, and the type of target—such as naval, air, or ground force; command and control center; line of communication; air defense (including SAM); surface-to-surface missile; early warning radar; telecommunications element; or piece of the infrastructure.
5. The expected presence of threats to the ISR platforms, such as SAMs, and their impact on ISR platform orbit location.
6. Day versus night missions. IR or synthetic aperture radar (SAR), but not visible EO sensors, must be used when image collections are required at night.
7. The regional weather conditions, including clouds, rain, fog, or haze, all of which may bring about complete obstruction or attenuation of the signals detected by the ISR sensors.

During our modeling effort, generation of the collection deck for each ISR platform was accomplished taking into consideration factors 1 to 6, as is explained briefly in the following paragraphs and in more detail in the following subsections. Factor 7 will be introduced in a future version of the model.

The ISR assets modeled were those of the existing ISR force; no attempt was undertaken to model future ISR capabilities.[4] Moreover, the composition and employment of the modeled ISR force was provided by information from the Terminal Fury 05 exercise, which included both Air Force and Navy airborne assets.[5] Thus, matching of sensors to platforms was per-

---

[4] Such work has been the subject of previous RAND studies. We did, however, explore the utility of the Global Hawk replacing the U-2 asset.

[5] This exercise, which takes place in the PACOM area of responsibility, includes a fully manned AOC with an ISRD.

formed using common and classified knowledge on Air Force and Navy airborne ISR capabilities. The platform orbits were generated as described below using orbit polygonal envelopes, which were also taken from the Terminal Fury exercise. We assumed that such orbits were adequate to avoid overflights of known threat zones, that is, that factor 5 above was taken care of by the designers of the Terminal Fury exercise.

Three sets of candidate targets were analyzed. The first set comprised 429 targets, on which information was to be collected during the selected day of the Terminal Fury exercise. The second set, containing 251 additional targets, originated from the master target list for the scenario. The last set consisted of 337 more targets from the Terminal Fury 03 exercise. Each target was assigned a unique target number or label, as well as a target category—namely, naval, air, or ground force; command and control center; line of communication; air defense; surface-to-surface missile; early warning radar; telecommunications; infrastructure; or directed search area.

Generation of the collection deck proceeded by "prefiltering" the target sets as follows. For each ISR platform/sensor combination and for each orbit point associated with that platform, all the candidate targets were tested for a certain number of criteria, laid down according to factors 2 (blue ISR capabilities) and 3 (geometrical considerations). Each target that became a candidate for collection for a particular orbit point was tagged with its unique target number or label for identification purposes, its assigned target category, the range from that target to the specified orbit point, the calculated National Imagery Interpretability Rating Scale (NIIRS)[6] value (for IMINT) and the squint angle with respect to the direction of flight of the ISR platform. The targets that were selected using the prefiltering process went to the next round of selection, based on PIR and operational and tactical ISR objectives, target characteristics, and whether the mission occurred during the day or at night—that is, factors 1, 4, and 6. In the following two subsections, we state and explain the criteria established for testing the targets during the prefiltering process. The next sub-subsection emphasizes the criteria set according to factor 2, blue ISR capabilities. The subsequent subsection elaborates further on the criteria that refer to factor 3, geometrical considerations.

Sensors. Imagery (IMINT) sensors used by the current ISR workforce can be of three types, namely, EO, IR, and SAR. The first two are passive sensors that collect images either in the visible (EO) or infrared (IR) portions of the electromagnetic spectrum. They are considered passive, since they rely on external sources of energy to illuminate the target—typically, direct or scattered sunlight for EO and thermal radiation from the target itself for IR. SAR is an active sensor that emits a radar signal toward the target, collects the returned radar echo, and analyzes it to create an image. SIGINT sensors can be of two types, that is, COMINT and ELINT; they involve specialized receivers and sophisticated processing electronics to analyze signals emitted by the targets in the form of telecommunication or radar emissions. Finally, GMTIs are capable of detecting moving vehicles on the ground against the background of possibly stronger radar returns from stationary objects (or clutter). During the current modeling effort, IMINT and SIGINT sensors were considered, but work on GMTI was delayed until the following study phase.

The geometrical and orbital considerations associated with the current ISR force of IMINT and ELINT sensors are described in Appendix B of this report. These sensors have different limitations that have to be carefully considered during the process of assigning platforms

---

[6] The NIIRS is discussed in more detail in Chapter Five.

and sensors to collect information on specific targets. In this sub-subsection, we focus on the impact that the virtues and limitations of such sensors have on the generation of the collection deck. One of the most important limitations relates to the maximum range—from platform to target—that can be achieved when using a particular sensor. The physical factors at the root of each limitation turn out to differ substantially from one type of sensor to another.

The predominant factor determining the maximum range for EO and IR IMINT sensors is degradation of image quality with distance. The quality of an image depends on image resolution, that is, the ability to resolve or distinguish two different but closely located objects. For these passive sensors, the farther the ground target is from the platform, the larger the footprint on the ground and the lower the image resolution. Below, we explain how one of the best criteria to measure sensor image resolution is by means of the associated NIIRS value. We also explain how the NIIRS value of a sensor is calculated from knowledge of the range and other sensor characteristics via the sensor NIIRS equation or via interpolation from measured data.

In addition to the NIIRS value associated with the sensor, each target can be related to three target NIIRS values, that is, the ones required for detection, for classification, and for identification of the target. Extensive tables of target NIIRS values have been generated for many kinds of targets; one example is shown in Table 4.1. The criterion on whether the sensor is capable of detecting, classifying, and/or identifying a particular target is simply that the calculated sensor NIIRS value be equal to or higher than the target NIIRS value needed to detect, classify, and/or identify such a target. The target NIIRS varies from 1 to 9, with the lower numbers associated with detection of large facilities and the higher numbers needed to identify small targets. For example, a NIIRS value of 1 is required to detect a port, whereas a NIIRS value of 6 is needed to identify a large ship.

As part of the prefiltering process applied to EO and IR collections, we calculated the sensor NIIRS value for each target–orbit point pair. Then we used two criteria for target selection; namely, the calculated sensor NIIRS value had to be (1) greater than or equal to 1 or

Table 4.1
NIIRS Value for SAR Against Several Target Types

	Required NIIRS for SAR Probability		
Target Type	**Detect**	**Classify**	**Identify**
Bridge	1	1	2
Large building	2	2	3
Large ship	2	3	6
Large aircraft	2	3	4
Patrol ship	3	3	5
Port	1	1	2
Petroleum, oil, and lubricants pump station	2	2	3
Petroleum, oil, and lubricants storage	3	3	5
Truck column	3	4	6

(2) greater than or equal to 3. Clearly, the former criterion allowed large targets, such as ports and large groups of vehicles, to be selected for the next step in the process of generating the collection deck.

Because of the physical mechanisms underlying SAR operation, the resolution of SAR images does not degrade with increasing the platform-to-target range.[7] However, the quality of a SAR image tends to be poor at low grazing angles. Accordingly, the primary factor limiting the maximum range achievable with a SAR IMINT sensor is degradation of image quality with decreasing grazing angle. Unfortunately, we were not able to find sound quantitative guidance on the degradation of SAR NIIRS values with decreasing grazing angle and were compelled to apply a pass/no pass criterion based on a selected minimum grazing angle. Two minimum grazing angles were chosen; namely, 3 and 6 degrees; the former is adequate to detect some kinds of targets (a building, a tower, or an armored brigade) but higher angles are needed for smaller targets such as vehicles, especially for classification and identification.

An important effect on SAR imaging is the bending of radar signals due to the stratified refractive index of the atmosphere, or atmospheric refraction. This effect was considered when calculating the maximum range achievable with the selected minimum grazing angle by assuming that the earth's radius is multiplied by 4/3.

In addition to examining whether the target-to-platform range exceeded the maximum sensor range, as previously defined according to the type of IMINT sensor, other tests were applied to the data as part of the prefiltering process. For example, restrictions concerning the sensor's depression or squint angles with respect to the direction of flight could apply. These restrictions are typically the result of mechanical limitations of the sensor assembly when mounted on the airborne platform.

The set of tests applied to the candidate EO and IR collections comprises the following:

- The platform-to-target range was compared with the minimum sensor range, as specified in the sensor's source data document (generally classified).
- The grazing angle at the target's position had to be strictly positive to take into account the earth's curvature, otherwise the sensor would be "seeing through earth."
- The calculated sensor's NIIRS value had to be larger than or equal to a specified number—either 3 or 1, as already stated. We examined both cases. A sensor's NIIRS value was calculated using a polynomial approximation to classified measured data, as explained below.
- The sensor's squint angle could not be greater than 10 degrees in absolute value. Squint angle is the angle between the platform's broadside line and the projection of the LOS vector onto the sensor's (or platform's) local horizontal plane.[8]
- LOS calculations at the target's position were performed to assess whether local terrain obstructions hindered the view from the platform.

[7] This assumes that there is sufficient power-aperture to not degrade with range.

[8] The sensor in question can be rotated in elevation, all the way to the nadir, though the nadir view is a recent fix to the software to avoid oversampling at the (relatively) short distance to the ground. The small squint angle range given to the sensor allows for capturing the pilot's ability to periodically maneuver the aircraft to use the sensor.

Concerning SAR, the following tests were applied to the candidate collections:

- Whether the platform-to-target range exceeded the minimum sensor range, as specified in the sensor's data document.
- Whether the platform-to-target range was smaller than the maximum sensor range, which was calculated from the selected minimum grazing angle, as described below. As already stated, two minimum grazing angles were used, 3 and 6 degrees.
- Whether the depression angle at the airborne platform's position was smaller than a maximum depression angle of 70 degrees. The depression angle is the angle between the LOS vector and the sensor's (or platform's) local horizontal plane.[9]
- Whether the squint angle at the airborne platform's position was smaller than a maximum squint angle (from the sensor's classified source).
- Whether the local topography did not obstruct the platform-to-target LOS based on digital terrain elevation data.

ELINT is an invaluable resource when searching for targets that emit electromagnetic signals but whose locations are not known. In a sense, the process of tasking ELINT platform/sensor combinations is simpler than for IMINT. ELINT sensors are typically assigned to collect signals of prespecified frequencies and waveforms. They thus collect *any* signal meeting those characteristics and that falls within the sensor's field of view, that is, the ELINT sensor operates similar to a "vacuum cleaner" of target emissions.

The following tests were applied to the candidate target collections for ELINT:

- We compared the target-to-sensor range with the maximum range as specified for the sensor.
- The grazing angle had to be a strictly positive number.
- The squint angle of the target with respect to the ISR platform had to be within one-half of the azimuth coverage, or azimuth beam width, specified for that ELINT sensor (where data were available).
- Just as for EO, IR, and SAR sensors, unobstructed LOS had to exist between the platform and target for a given platform location on orbit and assumed target location.

We tabulated the results from applying the prefiltering process to two target sets, as shown in Table 4.2 for EO/IR collections and in Table 4.3 for SAR collections. Each "hit" counts as a target being selected as a candidate collection for a specific orbit point and for a particular platform and sensor. In addition to the number of hits, the tables show the number of targets—including red and green targets—captured by the prefiltering process per orbit/platform/sensor combination. As expected, the number of selected targets is higher when the EO/IR sensor NIIRS value is compared with a lower NIIRS value (1 instead of 3), and

[9] The range resolution of the SAR becomes degraded as the depression angle of the SAR increases, and so one generally does not operate at depression angles greater than 70 degrees. In the case of mechanically scanned antennas, it is also possible that the gimbals are designed with this restriction. In the case of electronically scanned antennas, one would set the bore sight (if it is fixed) to point at maximum range, because there are antenna losses as one scans off bore sight. So it is also possible that there is a maximum depression angle associated with an electronically scanned array or active electronically scanned array. In any event, for one reason or another, unless there is information to the contrary, it is customary to assume a maximum depression angle of 65 or 70 degrees. For an illustration of the depression angle and other geometric references, please see Appendix B.

Table 4.2
Results from Prefiltering Applied to Two Target Sets, EO/IR Collections

10/13/05	NIIRS ≥ 3		NIIRS ≥ 1		Ratio of the number of targets in NIIRS 1 to the number in NIIRS 3
Target set 1	**Number of hits**	Number of targets	**Number of hits**	Number of targets	
EO_U2_R	1131	199 113 red, 86 green	2887	375 274 red, 101 green	1.9
IR_U2_R	317	80 5 red, 75 green	2767	361 262 red, 99 green	4.5
EO_U2_S	453	78 8 red, 70 green	1229	208 115 red, 93 green	2.7
IR_U2_S	229	44 2 red, 42 green	1027	177 91 red, 86 green	4.0
Target set 2					
EO_U2_R	149	40 13 red, 27 green	434	90 54 red, 36 green	2.3
IR_U2_R	86	25 3 red, 22 green	402	83 47 red, 36 green	3.3
EO_U2_S	208	30 6 red, 24 green	575	113 80 red, 33 green	3.8
IR_U2_S	133	21 3 red, 18 green	498	86 54 red, 32 green	4.1

RAND *TR459-Table 4.2*

Table 4.3
Results from Prefiltering Applied to Two Target Sets, SAR Collections

	gamma ≥ 6 degrees		gamma ≥ 3 degrees		Ratio of the number of targets in gamma 3 to the number in gamma 6
Target set 1	**Number of hits**	Number of targets	**Number of hits**	Number of targets	
SAR_U2_R	1479	122 13 red, 109 green	8135	329 219 red, 110 green	2.7
SAR_U2_S	797	56 2 red, 54 green	2106	129 42 red, 87 green	2.3
SAR_JSTARS_S	0	0	308	43 2 red, 41 green	—
Target set 2					
SAR_U2_R	389	50 6 red, 44 green	1072	77 32 red, 45 green	1.5
SAR_U2_S	399	25 4 red, 21 green	1026	62 31 red, 31 green	2.5
SAR_JSTARS_S	6	1 green	191	22 3 red, 19 green	22.0

RAND *TR459-Table 4.3*

when the grazing angle of the target with respect to the SAR sensor (or gamma) is compared with a lower angle (3 instead of 6 degrees). The last column in both tables lists the ratio of targets filtered with NIIRS 1 (gamma 3) criteria to those with NIIRS 3 (gamma 6).

Orbits and Geometrical Considerations

Criteria to prefilter targets according to orbits and geometrical considerations must be established. To do so, a survey of different representations of the earth is first conducted, and selection of a model that is appropriate for our goals of tasking ISR assets, in general, and generating collection decks, in particular, is necessary. Details are presented in Appendix B. The important subject of calculating angles of grazing, depression, and squint is also addressed, as well as the procedure followed to determine whether free LOS is available between a target and an orbit point.

Building Platform and Sensor Collection Decks

The process just described takes as inputs the entire set of potential collection targets and platform location as a function of time and sensor performance parameters. It outputs a list of filtered collections that can, physically, be collected by the sensor at the specified time. Each line of this list includes the target name and location, its category, the time at which it can be collected, and the NIIRS value (if imagery) of this collection. Note that individual targets can have multiple occurrences if they can be collected at multiple times. Since the time step used is one minute and the collection platform does not move a large distance in this time, multiple occurrences for each target are fairly common. The next step in creating our collection decks is to take this list and specify exactly what, out of all the possible choices, should be collected at each time step. This requires prioritization and depends on the collection strategy that is employed.

Collection Strategies

In reality, and in the modeling we have done, the strategies for prioritizing intelligence collection depend on a number of factors. We examined two potential collection frameworks, one based around the commander's prioritized information requirements and one using a utility ranking of ISR tasks. The latter one is based on work previously mentioned above. After discussing these two prioritization schemes, we walk through the final steps of building the collection plan, including optimizing imagery quality and allocating ISR resources.[10]

Note that this discussion will ignore several real-world complicating factors. First, we assume that collection managers have the freedom to choose the best set of collection targets to achieve the commander's objectives. In reality, some collection requirements will be mandated from above, and some fraction of the collection resources might be devoted to tasks outside the collection manager's control. These limitations could be easily accommodated in the methodology described below. Second, as indicated above, we concentrate here on airborne collection assets. In the real world, national systems and other collectors should satisfy some requirements. A simple method to account for them would be to take the targets that cannot be serviced by airborne platforms and assign them for national collection, but this may not make best use of either system.

Prioritized Information Requirements

The commander's PIRs are strategic-level intelligence needs that are often formulated as questions. For example, what is the enemy's disposition of theater ballistic missiles? These broad

---

[10] The IMINT collection prioritization process does not account for PED. While the second model, which is discussed in the next chapter, allows for analysis of the PED process, this study's scope does not explore this important topic. Future research could do so.

intelligence requirements are satisfied with specific information on particular enemy systems (in this example, theater ballistic missiles). Ultimately, answering the PIRs is accomplished by servicing a list of individual targets with a variety of sensors. The list of PIRs, and their relative priorities, forms the basis of the strategy for each collection deck. Note that the PIRs used to guide the collection plan can be at the JTF level or down at the individual component level.

The first step is to associate a list of target categories with each PIR using the EEIs and observables for each, along with the wording of the PIR itself. These categories include not only the target type, but also the side (a red forces radar transmitter as opposed to one on the green side) and range or depth. This latter is important because collection requirements closer to the forward line of troops (FLOT), for instance, might be more important than ones farther away.

Once each PIR has been associated with all of the target categories necessary to fully satisfy it (our current work has seven PIRs and 48 target categories), we next assign a detect, classify, or identify requirement to each category using the PIRs, EEIs, and observables. For example, if the PIR asks for occupancy checking of ports, detecting ships could be sufficient, while locating a SAM site might necessitate identification of a vehicle, assuming it is a mobile SAM. Each of the three requirements specifies different needs for the NIIRS imagery level or the type of SIGINT (ELINT, COMINT externals, COMINT internals, specific emitter identification, etc.) necessary.

Next, the list of possible collection targets is screened against the NIIRS and ELINT requirements just discussed and sorted by priority (collections not needed to satisfy any of the PIRs are discarded). For instance, if NIIRS level 6 is required of all red SAM sites within 100 km of the FLOT, all potential collections of this type with NIIRS values below 6 are excluded from further consideration.[11] The remaining set of potential collections is then sorted by the priority of the PIR that each collection supports. In the case of target types that support multiple PIRs, we use the highest-priority PIR for determining the priority of that target type. It may be worth considering for future work giving these targets some type of even higher priority, since, presumably, collections that support multiple PIRs increase the efficiency of the ISR system and so are preferred.

The final input necessary when using a PIR-based collection scheme is to assign a percentage of collection resources to each PIR. For imagery, this would be a number of imagery slots or for ELINT a percentage of time spent scanning the relevant frequencies. Typically, the highest priorities receive the highest percentage of resources. If there are not enough potential targets to collect with a given priority, remaining resources are rolled over to the next highest priority.

Utility Prioritization

A second methodology for prioritizing collection requirements is detailed in Chapter Three. Here, we utilize a strategies-to-tasks framework, using a weighting scheme applied to the campaign objectives and ISR tasks to generate a utility score for each potential collection target. Since a higher utility score indicates a greater contribution to accomplishing objectives, it

---

[11] Note that this gives a hard limit on lower-quality collections. It could be argued that, if a less-than-desired NIIRS level (or SIGINT type) was all that was available for a high-priority target, it should still be collected. However, determining a "corrected" priority for these less-useful collections would be problematic and so was omitted for this first iteration of the model.

also indicates a higher priority for collection. Whereas the previous, PIR-oriented methodology used the relative ranking of the PIRs to generate priority, this scheme uses those utility scores.

From the collection planning perspective, the only substantive difference is that there might be around one dozen priority levels (the number of PIRs) for the first scheme, but hundreds (the number of strategic objectives times the number of operational objectives times the number of ISR tasks) of different priorities for the utility methodology. While this number may seem like a disadvantage—since it precludes using a simple allocation of ISR resources to PIRs—it is actually an advantage in that we can avoid having to make this decision altogether. With hundreds of targets potentially associated with each PIR, and hence each priority level, it is necessary to use a percentage scheme to allow lower-priority targets to be collected at all. With a utility framework, in which each target or small group of targets will have unique priorities, we can simply start with the highest-priority targets and work our way down the list as far as possible until all of the collection resources are exhausted.

Allocation of Resources

Whether using the PIR or utility framework, we have now attached a priority to each potential collection target in our master list. Our next main objective is to take this list of potentially hundreds of possible collections and build a deck of collections that specify what target will be collected at what time by a particular sensor. Additionally, we must also specify when collections *should not* be planned in order to allow for flexibility for ad hoc requests and other dynamic requirements. Of course, a truly flexible plan allows for planned collections to be displaced by higher-priority dynamic targets, but having a well-structured and transparent plan prior to mission execution is critical to making smart decisions about these emergent targets.

Two more factors must be incorporated before building the final collection plan. First, all other things being equal, we prefer higher-quality collections over lower ones. Although we have already excluded those that did not meet the requirements, for those that exceed the requirement, we simply want to pick the highest NIIRS value or most preferred SIGINT type. Since we know the NIIRS level of each potential collection, it is a simple matter to sort within each priority level. The second factor to include is opportunity. For some targets, there may be a small time window in which collections can be made because of orbit geometry or external campaign events (collection just prior to a strike, for instance). For each potential target, we can calculate how many collection opportunities there are and preferentially, within each priority, choose the ones that have the fewest opportunities. Thus, the overall ranking scheme for potential collections is as follows:

1. *Priority.* Higher-priority targets will always be collected over lower ones.
2. *Opportunities.* Within each priority, targets with fewer collection opportunities will be collected before those with more.
3. *Quality.* Within the previous two constraints, higher-quality imagery (e.g., higher NIIRS values) will be collected over lower-quality imagery.

With all of the potential collections filtered and sorted as just described, building the actual collection plan is quite simple. Beginning with the highest-priority collection target with the fewest opportunities, we look up all of the possible times of collection. The time at which the highest NIIRS level (for imagery) can be collected is chosen and placed on the collection deck. If this target requires revisits at a regular interval, those revisits are also placed on the deck relative to the first one. We then move down the list to place other targets with the same priority but more opportunities. The time required for each collection determines how many may be placed at each one-minute time step. For example, we assume that a SAR image takes approximately one minute and that five EO/IR images could be taken per minute. If we are using a PIR-based framework, this process continues until we have placed the allowable number of collections (set by the percentage of total possible collections) for this priority. The overall number of collections is also limited to a user-input number to leave room on the deck for possible ad hoc targets. To ensure that ad hoc slots are spaced relatively evenly throughout the mission, we additionally allow the user to specify how many consecutive collections are allowed.

Future Work

At this point in the tool's development, there are several improvements to make that would add to its fidelity and utility, in addition to the features already mentioned. First, an individual collection deck is created for each sensor on each platform without reference to any other, which raises two obvious issues. If there are multiple sensors on board a single aircraft that must share the duty cycle—for example, SAR and GMTI—this limitation, if ignored, could lead to overstating the capability of the system. Offline analysis could be performed to give each sensor some fraction of the total duty cycle, but doing so would suboptimize collections, and different collection needs could require vastly different uses of the two sensors. Also, this implies that there is no coordination between multiple aircraft airborne at the same time—they may all attempt to collect the same targets. Again, preprocessing the list of potential collections to divide up the potential targets among possible collectors could help with deconfliction and coordination, but it may not be obvious outside the collection planning process itself what system or sensor is best to employ. The tool may need modification to allocate resources across sensors and platforms instead of just within them.

The inclusion of SIGINT collection planning could be improved in the tool as well. Resource allocation is currently modeled as conducting a certain number of collections per time step, but this framework does not truly capture SIGINT collection methods. Although the time spent scanning in a certain frequency range is one resource that must be considered, another could be analyst or linguist time. Similarly, SIGINT systems can collect many unplanned targets during the conduct of their missions. It is important to account for this additional "score." Also, there is not really a measure of collection quality that is comparable to NIIRS level for imagery. The collection planner may prefer certain types of collections or certain types of information about the signals collected, but the quality of a collection may be more a matter of luck (certain radar modes are observed or certain phrases are overheard) than anything else. As a result of all these issues, a specific version of the collection-management tool may be necessary to more appropriately model the SIGINT planning process.

CHAPTER FIVE

Collection Operations Model

Objectives and Approach

In the previous chapter, we saw how the CRT can be used to create a series of collection decks by platform, sensor, and orbit combination. We now turn our attention to the COM, which simulates the employment of ISR assets, with their assigned collection decks coming from the CRT, in a war scenario.

The objective of this modeling task is to provide the operational results that will permit evaluation of the costs and benefits of a given collection strategy. We also explore the results of exchanging sensor payloads on an existing platform and swapping a future Pacific asset for a current one. Specifically, we run the model flying a U-2 with a SAR sensor and later with an EO/IR sensor package. Note that the ELINT sensor package remains the same. Since PACAF is slated to receive the Global Hawk as a replacement for the U-2 platform, we also ran the scenario with the Global Hawk replacing the U-2.[1]

To examine the effect of using different collection strategies, we created an employment model that allows us to study a given strategy's outcome in the context of a scenario. The reason we use this framework is that numerous factors can affect the outcome of a given collection strategy. Factors include how the red side behaves during the conflict, for example—does it take an aggressive posture or does it practice camouflage, concealment, and deception techniques?

As mentioned in the previous chapter, the blue ISR assets have particular orbits, schedules, and system capabilities that affect the strategy's success. Also, the timing between one assigned blue forces asset and another is important, as well as the timing between those assets and the red systems' behaviors can affect the outcome. Some questions need to be considered. For example, when blue assets are assigned to the same region at similar times, can we take advantage of cross-cueing or fusing timely sensor data? And, in the latter case, do we fly collection assets during the day, with red forces coming out of hiding and emitting only at night? Things like weather and terrain masking can degrade system capabilities, too.

[1] When replacing the U-2 with the Global Hawk, sensor capabilities change, as well as the operating altitude and endurance of the platform. We did have the Global Hawk fly the same preplanned orbits as the U-2. Results are classified and, therefore, omitted from this report.

Model Description

The COM employs the blue ISR assets to find, fix, track, and target red forces' behavior. Besides the collection decks from the CRT, asset characteristics and the behavior and characteristics of red systems are fed into the model. Information on the platform's speed, endurance, and altitude are necessary to capture the asset's capability to collect information. The sensor's field of regard, resolution, and range tell us the quality of that information. The behavior patterns of various red targets are scripted into the model so that we can see how well different collection strategies do against them. Collection requirements or targets are tracked through the ISR portion of the kill chain to see whether they meet a given collection goal. For example, if the commander is interested in knowing only where large armed forces are gathered, we need to find, fix, and track them, but we do not need to target them. The model output is measures of performance regarding how well the particular strategy meets the prescribed collection goals. Effectiveness measures speak to how well the strategy meets the commander's overarching campaign objectives.

The operational flow within the COM is illustrated in Figure 5.1. For a given platform and sensor combination, there is a series of operational checks to determine whether a sensor can "see" a target. These checks include the following:

- Is the asset within the maximum detection range of the sensor it is employing?
- Does the sensor have LOS to the target (i.e., is it terrain masked)?
- Is the sensor within its squint angle limitation (if it has one)?
- For a SAR image, does the sensor have sufficient grazing angle?

These operational checks are listed for platform A with sensor B in Figure 5.1.

If these operational criteria are satisfied, the capability of the sensor to observe the red system is examined. For imagery collections, the capability is based on the NIIRS system. For ELINT collections, there is a series of calculations performed, culminating in a probability of detection per scan cycle. The ultimate question that is addressed is whether the given sensor meets the operational requirements to observe the red forces' behavior. The details of the sensor modeling for this analysis will be discussed shortly.

If all criteria are met, the information about an observation is passed to headquarters, thus facilitating retasking of additional assets to gain more information as necessary. Headquarters also operates as a central repository of information on targets. Delays associated with the processing, exploitation, and dissemination of the information are represented in the model and will be presented later in this section. We now lay out the various red forces' systems and describe how they are characterized.

Red Systems' Behaviors

A typical red country may have a variety of assets that would fall under two categories: fixed facilities and mobile systems. Fixed facilities would include command and control headquarters; infrastructure buildings; lines of communications (i.e., roads, ports, and railways); airfields; and petroleum, oil, and lubricants facilities. Mobile systems are often associated with fixed facilities; these include air defense systems (e.g., SAMs and early warning (EW) radars), airplanes, invasion forces, navy ships, and missile battalions. Another mobile target warranting future research is individuals.

Figure 5.1
Operational Flow Within the Collection Operations Model

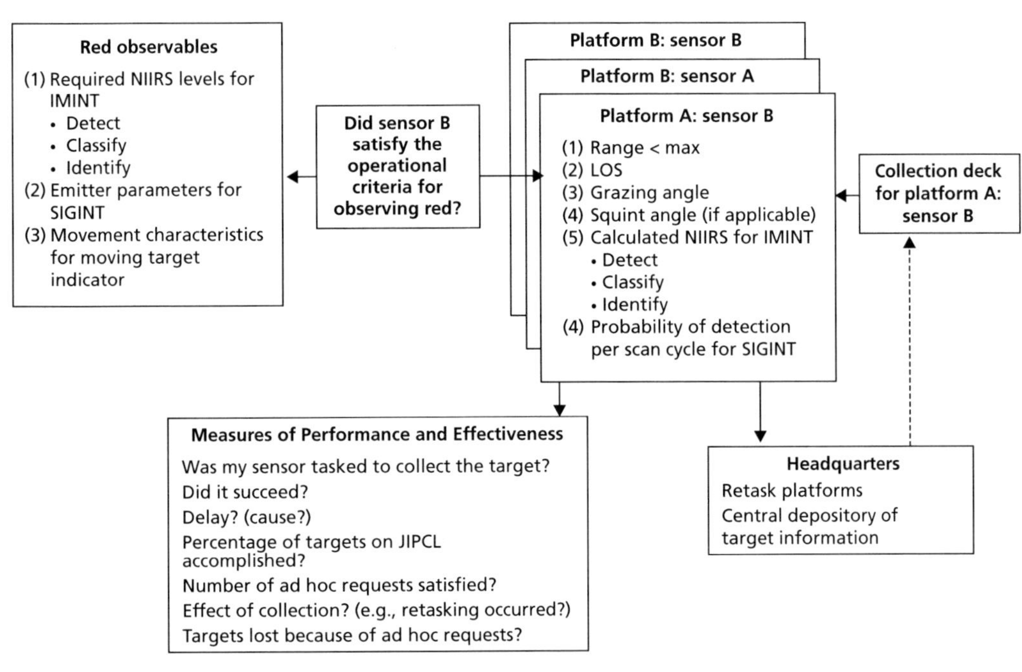

RAND *TR459-5.1*

The fixed facilities tend to be conducive for imagery collections or SIGINT gathering. Images provide information for targeting (e.g., coordinate mensurations from an optical image) and for occupancy checking (e.g., an army battalion leaving a garrison location). SIGINT collections on fixed facilities may also tell us whether the facility is currently occupied and perhaps by what type of personnel. For example, lower-echelon forces may tend to use particular emitters, while leadership (command and control) may use another type.

At the start of a campaign, mobile targets may be monitored at fixed locations if they have not yet dispersed and gone into hiding. Later, mobile targets need to be tracked when moving with a moving target indicator (MTI) radar. They may also emit signals enabling an ELINT collection.

For the initial scenario implemented, we characterize mobile surface-to-surface missiles as having aggressive behavior. They hide in locations unknown and periodically move, set up, launch a missile, move, and hide again. The blue forces' ability to see them occurs only when red forces first choose to move from hiding, and the time to catch them before launching a missile is minimal.

The air defense network is multilayered. EW radars provide long-range warning against blue air vehicles trying to penetrate red airspace or standing off its territory. In the scenario, EW radars are emitting at fixed locations. However, there are so many that, for a given region, only one at a time needs to emit to cover the region. Therefore, radars in the same region take turns emitting every two hours to provide complete coverage over the area of interest. SAM radars do not need to emit because they are networked into the EW system. SAM emissions would occur only when engaging a blue aircraft within range. While SAMs do not emit, they do not hide either. Our sensors, if directed to the correct location, would be able to image them.

Red invasion forces include army and navy assets. In the scenario, red forces are on the move as they invade a green country. Red forces congregate in particular locations, such as the FLOT and in staging grounds within the red country. Navy assets move from red ports across waterways toward the green country they intend to invade.

As mentioned previously, the red systems' behaviors are scripted in the scenario. The modeling environment provides the flexibility to have blue forces and red forces react to their perceptions of the other forces' actions. This flexibility will most likely be taken advantage of in the future. However, in the present scenario, we wanted to fix red's behavior and vary blue's collection strategies to collect information and evaluate these strategies. That way, model outcomes are based on blue's strategy solely and not on red's perceptions of or reactions to blue.

Blue Asset Characteristics

Just like red assets, blue ISR platforms are modeled explicitly in SEAS. With SEAS, we can create agents (or platforms) and assign characteristics to them. For example, airborne platforms must be given a speed, altitude, and range to operate. The time a platform departs from its air base and the duration it orbits on-station are explicitly set in the scenario and match the ISR synchronization matrix data provided from the military exercise. As a joint operation, all U.S. airborne ISR assets that flew in the military exercise were modeled in our scenario.[2]

Platforms are also assigned sensors and communication links. Communication links allow a platform to send information to a particular location, whether it is a deployable ground system, the Air Operations Center, or another platform. Long distances may separate a sensor from locations receiving the information, requiring intermediate relay stations, which can also be represented in SEAS, as well as any delays in data transmission.

Sensors assigned to platforms currently include EO, IR, SAR, and ELINT. Model development is continuing with the addition of COMINT and MTI sensors. We now describe the sensor representations in SEAS.

Imagery observations of a target are based on the NIIRS system as shown in Figure 5.2.[3] Before a NIIRS level is calculated, however, there is a series of operational specifications that must be met. First, the sensor must be within its grazing angle limitations. Next, it must be within its maximum range to the target. We must ensure that the target is not obscured by terrain, so a LOS check must be satisfied.

In our model, we have chosen as our primary measure of imaging sensor performance the NIIRS. The NIIRS is used within the intelligence community and by military collection managers in the military both as a means of rating the quality of existing imagery—e.g., for archival purposes—and for placing requirements on collectors in support of specific target reconnaissance. Recently, it has been adopted by the defense acquisition community and is frequently used for specifying the desired quality and performance of sensors to be developed and purchased by the Department of Defense.

The NIIRS scale, consisting of integers between 1 and 9 (and sometimes decimal fractions), measures the information potential of imagery for intelligence purposes, specifically,

---

[2] Platforms flown and collection results are presented in Lingel et al., unpublished.

[3] Atmospheric effects on sensors are presently not included. Cloud coverage is represented by a probability of coverage in a particular region based on average cloud coverage for that part of the world.

Figure 5.2
Imagery Observations Based on the NIIRS System

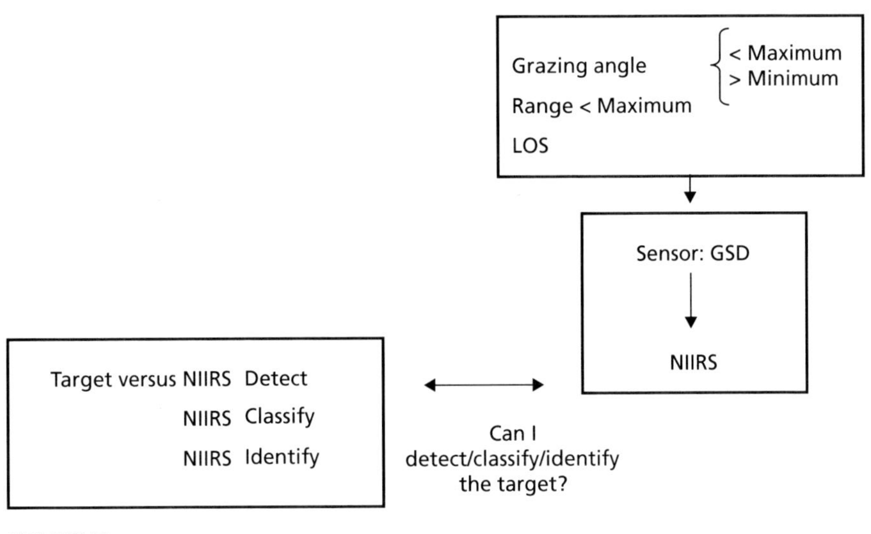

RAND *TR459-5.2*

the ability to detect, classify, and identify specific targets or target types under varying circumstances (e.g., in particular backgrounds, groupings, or installations). Thus, separate rating levels might be set for "detect medium-sized aircraft on tarmac" or "identify small fighter by type in hangar."

Clearly, in this functional form, the NIIRS can be applied only by (human) image analysts (IAs), and the relation between imagery products and sensor/processor designs is somewhat obscured. An unfortunate consequence of this situation in the past was that the government could not rigorously specify a desired sensor design in terms of NIIRS level, and the products of research and development efforts were often uneven. Efforts to objectify the scale, at least for optical sensors, have recently met with success, as we discuss subsequently. One of the early by-products of this success was the specification of Global Hawk's EO/IR sensors in terms of NIIRS value in the Global Hawk Advanced Concept Technology Demonstration program.

The interpretability of an image is attributable not only to the sensor's resolution, but also to the level of noise; the effects of atmospheric disturbances and sensor motion; and the impact on image quality of the optics, focal plane, and processing. Efforts to relate the NIIRS to a variety of objective sensor characteristics commenced in the 1960s. Some early efforts were based on the Modulation Transfer Function (MTF) and related parameters, on which sensor engineers typically rely. The MTF treats the image as composed of a distribution of spatial frequencies, and the sensor acts as a low-pass spatial filter, which passes the higher frequencies with decreasing amplitude. Since smaller objects correspond to higher frequencies, the MTF can be seen as a measure of the sensor's ability to provide small-feature information. Unfortunately, efforts to relate NIIRS directly to MTF, or to a related two-dimensional parameter (MTF area), were not satisfactory.[4]

---

[4] Possibly, this was because MTF underemphasized the importance of edge clarity in interpreting images.

Other criteria for image interpretability or utility have existed within the optics world for some time, including the well-known Johnson criterion, which relates the ability to detect, recognize, or identify targets based on the number of line pairs or pixels that can be resolved on the target. Resolving power is measured from the image of a standard display of groups of scaled parallel lines, in which each group consists of three lines separated by spaces of equal width. Since the criterion is sensitive to edge clarity, we would expect at least a reasonable degree of correlation with NIIRS. Perfect correlation should not be expected, however, since NIIRS is more target and background specific, and it can be responsive to the size of specific features that allow an IA to distinguish among target types or classes. Figure 5.3 shows a comparison of NIIRS with the Johnson criterion developed by the Aerospace Corporation. There is good agreement between the criteria with respect to detection and identification but not recognition, perhaps because it depends on features of intermediate scale that are highly variable.

A suitable general image-quality equation (GIQE), relating NIIRS to sensor parameters, was formally released in 1994. As modified by Leachtenauer et al., in 1997, the regression has the following form:

$$NIIRS = 10.251 - aLog_{10}GSM_{GM} + bLog_{10}RER_{GM} - 0.656H - 0.344\frac{G}{SNR},$$

where GSM is the geometric mean in inches of the two-dimensional GSD, RER is the relative edge response, H is the height overshoot caused by edge sharpening, G is the noise gain due to edge sharpening, and SNR is the signal-to-noise ratio. The parameters a and b have the values 3.32 and 1.559, respectively, when $RER \geq 0.9$, and 3.16 and 2.817, respectively, when $RER < 0.9$.

Figure 5.3
Comparison of Image Quality Criteria

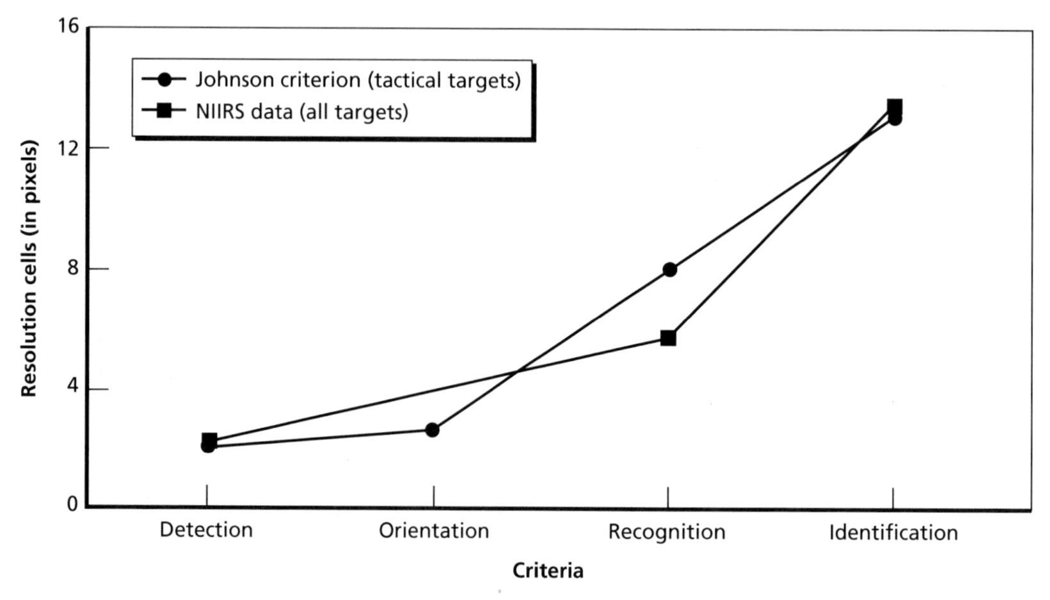

NOTE: Based on Gross, Andrews, and Hovanessian, 1991.
RAND *TR459-5.3*

Validation of the GIQE-predicted NIIRS against a population of 359 images that were rated by IAs resulted in a standard deviation error of 0.3 NIIRS value. Obviously, applying this equation is not for the faint of heart; a detailed knowledge of the optics, MTF, atmospheric effects, detectors, sensor motion, and processing is required. However, as a rule of thumb, if we substitute in mean values of the parameters (computed over the image population used to define the GIQE), the resulting equation is

$$NIIRS \approx c - 3.32 Log_{10} GSD_{GM},$$

with $c = 9.1915$. This formula comes close to approximating the mean Ground Sample Distance (GSD) values that have been informally associated with NIIRS ratings in the past. In our model, we use NIIRS values as a function of range whenever these data are provided by the contractor, program office, or other reliable source. When these data are not available, we use the contractor-specified resolution to compute GSD as a function of range and the above approximation for NIIRS. Thus far, we are assuming clear atmospheric conditions for optics and no rainfall for radars, but, in a later phase, NIIRS data will be modified to account for propagation degradation.[5]

In recent years, separate NIIRSs have been adopted for IR imagery, SAR imagery, and multispectral imagery, and a provisional NIIRS even exists for hyperspectral imagery. Unclassified versions of the NIIRSs for visible (both military and civil), IR, radar, and multispectral imagery are tabulated in Appendix A (Leachtenauer and Driggers, 2001). To date, there is no NIIRS for inverse SAR (ISAR) imagery. The ISAR NIIRS, which would be restricted, is a subject of ongoing research in our modeling effort.

A commonly used rule of thumb for IR NIIRS employs the preceding equation with $c = 9.82$. GIQEs for radar, multispectral, and hyperspectral imagery have not been published at this time. However, noting that NIIRS-7 functions are typically carried out at 1-ft resolution, we venture that taking $c = 10.5$ provides a reasonable rule of thumb for SARs operating where the signal-to-noise ratio (SNR) is high. As the SNR degrades at long range, this NIIRS estimated value will no longer be valid. For this reason, we have placed a premium on obtaining NIIRS data for SARs based on evaluation of real imagery by IAs.

One difficulty attending use of NIIRS is its incompleteness. Not every target category included in our campaign analysis appears in NIIRS. In discussions with analysts at the Defense Intelligence Agency, we were advised to extend the NIIRS by analogy—i.e., use similar NIIRSs for similar targets. We have done so but cannot claim any official acceptance of our choices.

Some general observations are in order concerning the dependence of NIIRS on range. NIIRS for SARs tend to be relatively insensitive to range until SNR effects come into play—because the synthetic aperture increases with range to preserve azimuthal resolution. NIIRS levels for EO are generally better than for IR, because of the poorer resolution at longer wavelength—whether resulting from diffraction-limited optics or the larger IR detectors on the

---

[5] In some of our earlier work, we have leaned toward the use of receiver operating characteristics (ROCs) to characterize sensor performance, rather than more subjective criteria such as NIIRS. An important difference between these approaches is that ROCs are intended to quantify false detections, while NIIRS is focused entirely on achieving image quality just sufficient to complete the task—i.e., detection. ROC curves have been developed in some instances (for forward-looking infrared radars and SARs) for detection by automatic target-recognition, but only for limited target sets and for specific target-recognition algorithms. For these reasons, they are not generally appropriate for this work.

focal plane. Some exceptions have been noted, e.g., because of inferior optical tolerances that affect the shorter visible wavelengths more than the IR. In clear air, the most significant factor in degrading optics performance relative to SAR is the increase in GSD with range, because optical systems have fixed apertures and consequently fixed angle resolution. An exceptional situation occurs near nadir, where the SAR's measurement of slant range yields poor resolution along the ground.

Although NIIRS provides useful guidance, assigning sensors based on this scale alone raises a number of troubling questions. The size of salient target features contributing to classification or identification is intrinsic to the NIIRS concept, but, practically speaking, this information is situation dependent. Choosing, for concreteness, the problem of searching for a headquarters building at a known military base, we note that the building might be distinguished from other structures on the base by reading its structural characteristics, the size of its parking lot, the presence of many communication antennas, the logo on the building, traffic analysis of vehicles on the base, a SIGINT intercept tied to the building, etc. Only the first of these is properly captured in the advertised NIIRS ratings for military headquarters, although the last one, which involves fusing IMINT with SIGINT, might be the most critical one if we desire to find an alternative or covert headquarters, the whereabouts of which are unknown. These considerations belong properly under the heading of "data fusion." The fusion methodology in our model is under development currently.

Targets in motion present a different set of issues. Vehicles moving farther than a pixel-width during the collection interval will cause blurring in both optical and radar images. While there are ways of compensating somewhat for blurring in both cases, the SAR image will generally suffer more degradation. First, the collection interval for SAR is typically orders of magnitude longer than for optics, so the blurring is more extreme. Second, the SAR processor utilizes Doppler—i.e., differences in radial velocity—to obtain azimuthal resolution, so movement will affect the entire image formation process. Thus, radial movers in SAR images are offset in position as well as blurred. Finally, since SAR image formation is phase coherent, it depends on compensating for relative motion between the target and sensor to within a fraction of a wavelength (~1 cm). ISAR techniques, which allow for shortening the collection interval when the vehicle is rotating or in a turn, can help under some circumstances, particularly at sea, where the rotation rates are rapid and do not depend on vehicle translation. Good results have also been achieved in classifying vehicles with range templating in a high-range-resolution (HRR) GMTI tracking mode. HRR has been shown to be useful in identifying features on vehicles that would aid in keeping targets in track (e.g., when close approaches in traffic threaten to confuse tracks). HRR performance results obtained by the Air Force Research Laboratory will be incorporated in the next phase of our model. Though HRR "images" may be obtained much faster than ISAR for land vehicles, on the ocean, ISAR imaging is superior both with respect to resolution and time for image capture.

Before we move on to a discussion of our ELINT model in SEAS, we must first address the calculation of the GSD for the various sensors. Recall that the GSD is the geometric mean of down-range, *DR,* and cross-range, *CR,* resolutions (in inches) or

$$GSD = \sqrt{DR \times CR}.$$

For a SAR sensor, down-range resolution is a function of the slant range, *SR*, and the grazing angle, γ, as follows:

$$DR = SR / \cos(\gamma).$$

For EO and IR sensors, the down-range resolution is the range (in inches) times the angular resolution, $\Delta\theta$ (in radians) divided by the sine of the grazing angle. The cross range is the range multiplied by the angular resolution. Therefore,

$$GSD = R \times \Delta\theta / \sqrt{\sin(\gamma)}.$$

The ELINT representation in SEAS is more complicated. The model is influenced by diverse factors pertaining to the scenario, the receiver design and operation, and the emitter characteristics and operations. In addition, there are stochastic elements, e.g., the likelihood that the ELINT antenna is pointing toward the emitter, with its receiver tuned to the emitter frequency, at the same time that the emitter's main beam is illuminating the ELINT aircraft. Figure 5.4 portrays the processes and features included in the model.

The scenario prescribes each emitter's location, operations, and characteristics, including field-of-regard or scan limits, frequency, waveform, beam width, scan time, effective radiated power (ERP), bandwidth, pulse repetition frequency, side-lobe level, on-off intervals, and emissions control discipline. Initially, with only EW radars included in the model, this parameter set was complete. The dataset for fire control or engagement radars requires additional specifications, including cueing and target reacquisition protocols, firing doctrine, and tracking update rates, much of which remains unavailable at this time.

Figure 5.4
ELINT Model Involves Many Factors

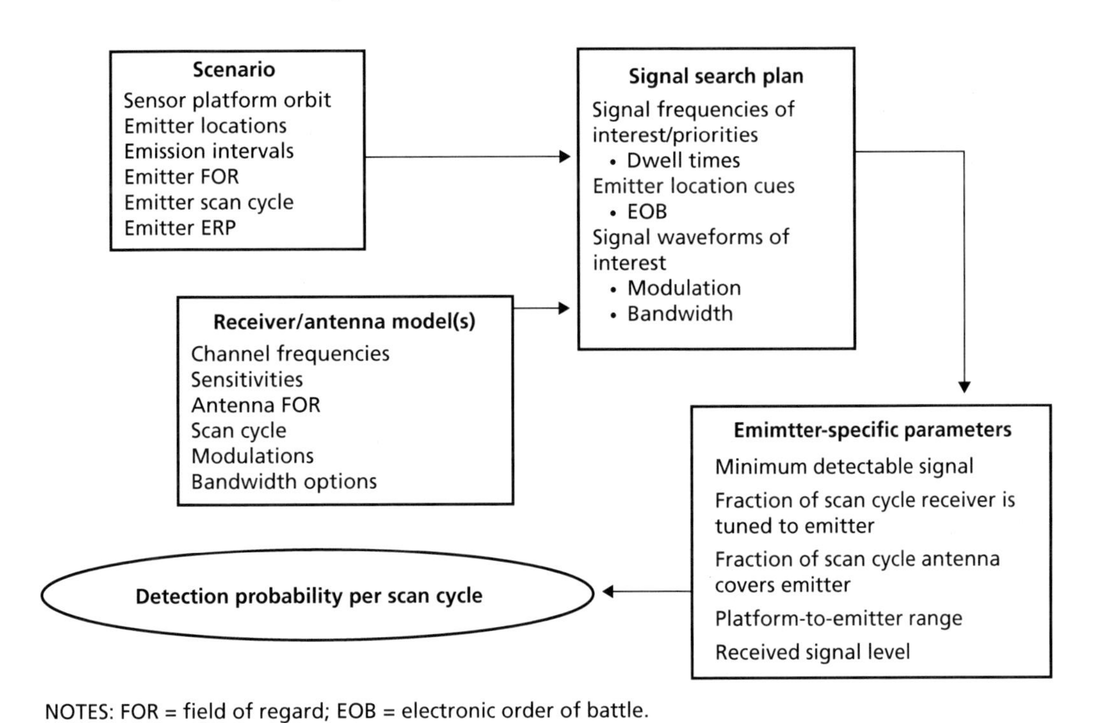

NOTES: FOR = field of regard; EOB = electronic order of battle.
RAND *TR459-5.4*

The modeling of COMINT externals follows a similar pattern to that shown in Figure 5.4. However, the modulations are typically more complex and diverse. Low ERP for some mobile transmitters can present severe challenges for interception, interference cancellation, and geolocation. In state-of-the-art COMINT receivers, sophisticated digital processing can play a significant role in extracting these signals, which is not yet included in the model.

The scenario prescribes the ELINT platform's orbit and, thus, establishes the range between the emitter and ELINT receiver at any time step. The propagation loss between the emitter and receiver is obtained from the range.

If the emitter's ERP and propagation loss are known, the only additional parameter required to compute the power incident on the receiver is the gain of the emitter antenna in the receiver's direction. We employ a simple approximation, replacing the gain pattern with a step function, so that the receiver is assumed to be illuminated with either the emitter's main lobe gain or average side lobe. With this simplification, the probability of main lobe or side lobe illumination depends on only the fraction of the receiver's scan cycle during which it is tuned to the emitter and the fraction of the emitter's antenna scan in which its beam is pointed at the receiver.

Subject to the availability of gain patterns for threat emitters, this model will be upgraded in a later phase of development. The upgrade would involve replacing the step function with a cumulative probability function describing the emitter's gain pattern.

Detection of the emitter's signal is assumed to occur if the incident power exceeds the sensitivity threshold of the receiver. Once again, we have chosen a step function approximation over the more precise employment of a receiver operating characteristic (ROC). (The latter would associate a probability of detection with values of the false alarm rate and signal-to-noise ratio.) Employing a ROC curve would require more detailed knowledge of the ELINT receiver designs than have been provided to us.

Most signal acquisition antennas on ELINT aircraft have fixed beams pointing broadside, or nearly so, and covering approximately a quadrant. These antennas have gain loss away from broadside, which is included in the model when the data are available. Using the emitter-receiver geometry specified in the scenario, the model computes the direction cosines from the receiver to the emitter to obtain the gain roll-off. There are typically some steerable pencil beam antennas, interferometric arrays, etc. available on ELINT aircraft as well, and it is planned to include them in the next phase.

The type of receiver currently employed in our model is a channelized scanning receiver. In general, a receiver of this kind can be programmed to operate flexibly and responsively in the anticipated threat environment. Thus, it is possible to scan with an intermediate frequency bandwidth (chosen from a small set of available intermediate frequency stages) that most closely matches the expected emitters in each frequency channel and to dwell at each frequency for a variable period that reflects the importance of each threat. The matching of bandwidths is desirable to optimize detection probability by excluding noise and interference external to the desired signal. In our discussions within the collection community, we have found that the degree to which operations are optimized can vary widely from careful weighting of dwell frequencies and intervals to uniform scanning through the full frequency spectrum. The latter tactic is resident in our model at this time, subject to more detailed guidance on our scenario options.

The receiver's single-pulse detection threshold varies with the frequency, waveform modulation and bandwidth, and the bandwidth of the channel in which the pulse or waveform sample is collected. Much of the frequency dependence is caused by the employment of different antennas for each frequency channel, and, of course, the gain of broadband antennas typically varies with frequency as well.

The probability of detecting a pulse increases if more than one pulse can be intercepted during the receiver dwell in the emitter's bandwidth. The number of captured and potentially detected pulses is equal to the dwell divided by the pulse repetition interval. The equations that determine whether a single pulse is detected are displayed in Figure 5.5.

The equations simply state that detection occurs when the received power exceeds the sensitivity threshold, taking into account the various factors noted above. The form of the noise term indicates that detection is assumed to be limited by thermal noise, which is not always the case, since there are other potentially dominant interference sources, including electromagnetic interference from other platform electronics, jammers, and the presence within LOS of a diverse population of military and civilian emitters.

Platform noise, if it dominates thermal noise, could lead to different performance levels for standard ELINT payloads when they are installed on different types of aircraft or when various modular payload mixes are loaded onto the same aircraft. High emitter density is more likely to affect ELINT receivers on high flyers such as Global Hawk, because of the larger population of emitters circumscribed within their extended LOSs. The negative effects of dense emissions can extend beyond the detection process and severely complicate the job of the digital processor in sorting, associating, and identifying pulse streams. The effects of nonthermal noise are not currently modeled.

Figure 5.5
ELINT Model Detection Equations

$$Sensitivity(dBW) = -RadomeLoss - AntennaGain + NoiseFigure + Threshold + kT_o + 10 \times Log_{10}(B_{Hz}) - AntennaRolloff$$

Threshold depends on desired probability of detection, P_d, and FAR

$$kT_o = -204 dBW/Hz$$

$$AntennaRolloff(dB) - 10Log_{10}(\cos^3 \theta)$$

θ = azimuth from bore sight

$$FluxDensity(dBW/m^2) = 10Log_{10}(4\pi/\lambda^2) + Sensitivity(dBW)$$

$$\lambda = wavelength(m) = speed\ of\ light \div frequency$$

$$FluxDensity(emitter) = \frac{EffectiveRadiatedPower}{(4\pi)Range^2}$$

Signal is detected if $FluxDensity(emitter) \geq FluxDensity(dBW/sq\ m)$

RAND *TR459-5.5*

MTI Description

The characterization of MTI in SEAS is based on blue's collection strategy, red targets' behaviors, and the allocation of available radar resources. Rather than developing a detailed tracking model within SEAS, we rely on off-line tracking analysis performed from previous RAND research to inform us on the necessary resources to do a range of possible MTI missions. Previous analysis provides resource requirements for MTI for a range of tasks.

Each MTI task necessitates a given revisit rate based on factors that include the target type, traffic density, and the environment. Tasks range from general situational awareness to tracking an individual vehicle in heavy traffic in an urban environment. Typical required revisit rates are listed in Table 5.1 for various tasks.

With multiple competing tracks to perform, decisions must be made among tasks when resources are 100 percent tasked. This situation forces the intelligence collector to make difficult tasking decisions on what targets to pursue and what ones to drop and requires an explicit blue collection strategy to guide the choice. In this way, we demand honest bookkeeping of MTI resources.

Blue Concepts of Operations for Collecting Red Information

All sensor characterizations are assembled into a series of blue CONOPSs for collecting information against red targets. The collection process may begin from foreknowledge of the scenario (IPB), such that an asset is tasked to look in a particular location based on previous red observed behavior. Or a cue from an intelligence source such as human intelligence or a national system may start the process. Once new information is gained, it is sent back to its associated command center. For example, a U-2 would send an image back to its Distributed Common Ground Station, and the data would be forwarded to the AOC. As information is passed, there is an associated delay due to processing, exploitation, and dissemination steps.

How this process is characterized in the model may be better understood through an illustration. Imagine a Special Operations Forces platoon at an observation post outside of a town. As a small convoy of suspect vehicles passes, the Special Operations Forces observe them and send information back to their command center describing what they saw. Since the Special Operations Forces are hiding and on foot, they cannot pursue the vehicles. They are also describing verbally what they see; so the location accuracy of the information is poor.

Once command center personnel receive the information, they must make a decision (a delay here) to act on it by retasking an airborne resource that happens to be in the area. In this example, a COMINT asset is there and listens to communications that the vehicle occupants are having with someone back in town. The COMINT collection provides positive identifica-

Table 5.1
MTI Tasks and Associated Revisit Rates

Type of Monitoring	GMTI Revisit Rate
Situational awareness	Double-digit minutes
Track large vehicle groups	Single-digit minutes
Track individual vehicles in low traffic	Double-digit seconds
Track individual vehicles in heavy traffic	Single-digit seconds

tion of a particular individual, but the location accuracy is poor, and COMINT is unable to persist in the area because of other taskings.

The COMINT information is sent to the COMINT command center and forwarded to another command center (more delays) until a platform with GMTI is tasked to track the vehicle group. The group is then tracked until some sort of endgame occurs. The entire CONOPS process with embedded delays due to PED and command and control decisions can be represented in the SEAS model.

Delays are based on current PED capabilities. Air Force intelligence officers were asked to provide a range of delays for each step given current processes. Future ranges based on automatic target-recognition capabilities or other resources could easily be implemented.

CHAPTER SIX

ISR Assessment from the Model

In the previous two chapters, we presented two models. The CRT created planned collection decks for sensors and prescribed platforms flying fixed orbit tracks. These planned collections are formed based on a given collection strategy. Two strategies were described, a PIR-based strategy and a utility-based strategies-to-tasks one. The collection decks are input for the COM to test the performance of a given strategy executed in a scenario.

Once the model is run, we must have a means to evaluate the performance and effectiveness of a particular strategy given the assets provided. Performance measures may be thought of as direct output from sensors in the model. For example, was the tasked sensor able to collect the requested information? What was the cause of any delay in collection?

Effectiveness measures may be harder to derive from the model because they judge the ultimate effect or outcome of a scenario based on how the ISR assets performed. For the first year of model development, we focused more on performance measures with the intention, in the near future, of expanding our output measures to include effectiveness. Ultimately, the goal is to learn what output measures should be monitored to gauge the success at a given effect.

Turning our attention to measures of performance, there are a number of questions to ask after the scenario is run. These may include the following:

- Was the sensor tasked to observe the event?
- Did the sensor observe the event or not?
- Are we tasking the right sensor against the right target?
- What was the delay in observation of the event?
- What percentage of the event was observed relative to ground truth?
- For MTI, how long did we keep the target in track and when during the event?
- What were the measuring errors, e.g., target location error or sensor error?

Higher-level questions consider the effect at the mission level or, in terms of timelines, the ATO day. They may provide insight regarding the applicability of a chosen collection strategy, such as the following:

- Does the collection deck accurately reflect priorities and guidance?
- What is the effectiveness of our airborne tracks? Could they be moved to improve effectiveness?
- Are ad hoc requests being given the proper priority? Are they preventing the satisfaction of PIRs? Should more flexibility be built into the preplanned collection deck?
- Are we accurately deconflicting resources?

- What was the ATO or mission outcome relative to the baseline case?
- Are collection elements using the most appropriate reporting vehicles in support of requirements? Are the vehicles and timelines sufficient to meet tasked objectives? Are real-time dissemination techniques conveying the desired depth of information?
- By target class, on which targets is information collected (how does this compare to the PIRs)?
- How many targets make it through the find, fix, track, and target portion of the kill chain, by target class?
- How many SAMs did blue perceive relative to ground truth?
- Are we scheduling and placing our airborne assets to coincide with windows of opportunity for collection, given current IPB? Are theater assets located for optimum geolocation and collaboration?
- Does our scheduling allow for improved cross-cueing and fusion of information?

Model runs validating the code and assessment process will be presented in a forthcoming volume. A PIR collection strategy, in which percentages of ISR assets are assigned toward PIRs, is examined. We also implement the strategies-to-tasks framework to create collection decks. Model output shows that the distribution of targets collected against varies between the two strategies. Unfortunately, the comparison between the two is an apple to an orange. At the model performance level, who is to say that one strategy is better than another one? The analysis must occur at a higher mission level. The broader question to address is, which strategy did better at meeting the commander's goals for the day? There is another wrinkle in this process that must be presented. The PIR collection strategy decks were based on the commander's PIR list for the day. The strategies-to-tasks framework decks originated from the commander's AOD for the day. These two documents should have similar priorities. However, they were not consistent. Therefore, for the preliminary results, we were not able to assess the strategies to the degree discussed in this chapter.

We did explore the effect of altitude, orbit location, and sensor capabilities on the performance within a given collection strategy. For example, we flew the Global Hawk as a replacement for the U-2. The Global Hawk, at a slightly lower altitude, was not able to see as far into enemy territory as was the U-2. We also saw the impact of altitude and orbit location on collecting ELINT on early warning radars.

Future model development will improve output information to better assess the employment of ISR assets. Development also includes the ability to cross-cue between sensors residing on the same platform or another platform, incorporating national assets into the force mix, and enabling COMINT and maritime radar capabilities.

The COM allows exploration of multiple factors. Not only is this methodology applicable to examining different collection strategies, but we can also study variations in the number of ad hoc slots provided for a given collection deck (i.e., sensor), examine the utility of new platforms or sensors, and evaluate the effects of different orbit locations. We may also analyze different strategies for retasking assets, consider the effects of PED delays, or appraise the synergy of multiple types of intelligence on a given target. Furthermore, we may look at the effect of a jamming environment on communicating information among assets and headquarters and explore the benefits of automatic target recognition to enable the collection of more information.

APPENDIX A

NIIRS Tables

Table A.1
Visible NIIRS, March 1994

Rating Level 0

Interpretability of the image is precluded by obscuration, degradation, or very poor resolution.

Rating Level 1

Detect a medium-sized port facility and/or distinguish between taxiways and runways at a large airfield.

Rating Level 2

Detect large hangars at airfields.

Detect large static radars (e.g., AN/FPS-85, COBRA DANE, PECHORA, HENHOUSE).

Detect military training areas.

Identify an SA-5 site based on road pattern and overall site configuration.

Detect large buildings at a naval facility (e.g., warehouses, construction hall).

Detect large buildings (e.g., hospitals, factories).

Rating Level 3

Identify the wing configuration (e.g., straight, swept, delta) of all large aircraft (e.g., 707, CONCORD, BEAR, BLACKJACK).

Identify radar and guidance areas at a SAM site by the configuration, mounds, and presence of concrete aprons.

Detect a helipad by the configuration and markings.

Detect the presence/absence of support vehicles at a mobile missile base.

Identify a large surface ship in port by type (e.g., cruiser, auxiliary ship, noncombatant/merchant).

Detect trains or strings of standard rolling stock on railroad tracks (not individual cars).

Rating Level 4

Identify all large fighters by type (e.g., FENCER, FOXBAT, F-15, F-14).

Detect the presence of large individual radar antennas (e.g., TALL KING).

Identify, by general type, tracked vehicles, field artillery, large river crossing equipment, wheeled vehicles when in groups.

Detect an open missile silo door.

Determine the shape of the bow (pointed or blunt/rounded) on a medium-sized submarine (e.g., ROMEO, HAN, Type 209, CHARLIE II, ECHO II, VICTOR II/III).

Identify individual tracks, rail pairs, control towers, switching points in rail yards.

Rating Level 5

Distinguish between a MIDAS and a CANDID by the presence of refueling equipment (e.g., pedestal and wing pod).

Identify radar as vehicle-mounted or trailer-mounted.

Identify, by type, deployed tactical SSM systems (e.g., FROG, SS-21, SCUD).

Distinguish between SS-25 mobile missile TEL and missile support vans (MSVs) in a known support base, when not covered by camouflage.

Identify TOP STEER or TOP SAIL air surveillance radar on KIROV-, SOVREMENNY-, KIEV-, SLAVA-, MOSKVA-, KARA-, or KRESTA-II-class vessels.

Identify individual rail cars by type (e.g., gondola, flat, box) and/or locomotives by type (e.g., steam, diesel).

Rating Level 6

Distinguish between models of small/medium helicopters (e.g., HELIX A from HELIX B from HELIX C, HIND D from HIND E, HAZE A from HAZE B from HAZE C).

Identify the shape of antennas on EW/GCI/ACQ radars as parabolic, parabolic with clipped corners, or rectangular.

Identify the spare tire on a medium-sized truck.

Distinguish between SA-6, SA-11, and SA-17 missile airframes.

Identify individual launcher covers (8) of vertically launched SA-N-6 on SLAV-class vessels.

Identify automobiles as sedans or station wagons.

Rating Level 7

Identify fitments and fairings on a fighter-sized aircraft (e.g., FULCRUM, FOXHOUND).

Identify ports, ladders, vents on electronic vans.

Detect the mount for antitank guided missiles (e.g., SAGGER on BMP-1).

Detect details of the silo door hinging mechanism on Type III-F, III-G, II-H launch silos and type III-X launch control silos.

Identify the individual tubes of the RBU on KIROV-, KARA-, KRIVAK-class vessels.

Identify individual rail ties.

Rating Level 8

Identify the rivet lines on bomber aircraft.

Detect horn-shaped and W-shaped antennas mounted atop BACK TRAP and BACKNET radars.

Identify a hand-held SAM (e.g., SA-7/14, REDEYE, STINGER).

Identify joints and welds on a TEL or TELAR.

Detect winch cables on deck-mounted cranes.

Identify windshield-wipers on a vehicle.

Rating Level 9

Differentiate cross-slot from single-slot heads on aircraft skin panel fasteners.

Identify small, light-toned ceramic insulators that connect wires of an antenna canopy.

Identify vehicle registration numbers (VRN) on trucks.

Identify screws and bolts on missile components.

Identify braid of ropes (3–5 inches in diameter).

Detect individual spikes in railroad ties.

Table A.2
IR NIIRS, April 1996

Rating Level 0

Interpretability of the imagery is precluded by obscuration, degradation, or very poor resolution.

Rating Level 1

Distinguish between runways and taxiways on the basis of size, configuration or pattern at a large airfield.

Detect a large (e.g., greater than 1 km^2) cleared area in dense forest.

Detect large ocean-going vessels (e.g., aircraft carrier, super-tanker, KIROV) in open water.

Detect large areas (e.g., greater than 1 km^2) of marsh/swamp.

Rating Level 2

Detect large aircraft (e.g., C-141, 707, BEAR, CANDID, CLASSIC).

Detect individual large buildings (e.g., hospitals, factories) in an urban area.

Distinguish between densely wooded, sparsely wooded and open fields.

Identify an SS-25 base by the pattern of buildings and roads.

Distinguish between naval and commercial port facilities based on type and configuration of large functional areas.

Rating Level 3

Distinguish between large (e.g., C-141, 707, BEAR, A-300 AIRBUS) and small aircraft (e.g., A-4, FISHBED, L-39).

Identify individual thermally active flues running between the boiler hall and smoke stacks at a thermal power plant.

Detect a large air warning radar site based on the presence of mounds, revetments, and security fencing.

Detect a driver training track at a ground forces garrison.

Identify individual functional areas (e.g., launch sites, electronics area, support area, missile handling area) of an SA-5 launch complex.

Distinguish between large (e.g., greater than 200m) freighters and tankers.

Rating Level 4

Identify the wing configuration of small fighter aircraft (e.g., FROGFOOT, F-16, FISHBED).

Detect a small (e.g., 50m^2) electrical transformer yard in an urban area.

Detect large (e.g., greater than 10m diameter) environmental domes at an electronics facility.

Detect individual thermally active vehicles in garrison.

Detect thermally active SS-25 MSVs in garrison.

Identify individual closed cargo hold hatches on large merchant ships.

Rating Level 5

Distinguish between single-tail (e.g., FLOGGER, F-16, TORNADO) and twin-tailed (e.g., F-15, FLANKER, FOXBAT) fighters.

Identify outdoor tennis courts.

Identify the metal lattice structure of large (e.g., approximately 75m) radio relay towers.

Detect armored vehicles in a revetment.

Detect a deployed transportable electronics tower (TET) at an SA-10 site.

Identify the stack shape (e.g., square, round, oval) on large (e.g., greater than 200m) merchant ships.

Rating Level 6

Detect wing-mounted stores (i.e., ASM, bombs) protruding from the wings of large bombers (e.g., B-52, BEAR, BADGER).

Identify individual thermally active engine vents atop diesel locomotives.

Distinguish between a FIX FOUR and FIX SIX site based on antenna pattern and spacing.

Distinguish between thermally active tanks and APCs.

Distinguish between a 2-rail and 4-rail SA-3 launcher.

Identify missile tube hatches on submarines.

Rating Level 7

Distinguish between ground attack and interceptor versions of the MIG-23 FLOGGER based on the shape of the nose.

Identify automobiles as sedans or station wagons.

Identify antenna dishes (less than 3m in diameter) on a radio relay tower.

Identify the missile transfer crane on an SA-6 transloader.

Distinguish between an SA-2/CSA-1 and a SCUD-B missile transporter when missiles are not loaded.

Detect mooring cleats or bollards on piers.

Rating Level 8

Identify the RAM airscoop on the dorsal spine of FISHBED J/K/L.

Identify limbs (e.g., arms, legs) on an individual.

Identify individual horizontal and vertical ribs on a radar antenna.

Detect closed hatches on a tank turret.

Distinguish between fuel and oxidizer multisystem propellant transporters based on twin or single fitments on the front of the semitrailer.

Identify individual posts and rails on deck edge life rails.

Rating Level 9

Identify access panels on fighter aircraft.

Identify cargo (e.g., shovels, rakes, ladders) in an open-bed, light-duty truck.

Distinguish between BIRDS EYE and BELL LACE antennas based on the presence or absence of small dipole elements.

Identify turret hatch hinges on armored vehicles.

Identify individual command guidance strip antennas on an SA-2/CSA-1 missile.

Identify individual rungs on bulkhead mounted ladder.

Table A.3
Radar NIIRS, August 1992

Rating Level 0

Interpretability of the imagery is precluded by obscuration, degradation, or very poor resolution.

Rating Level 1

Detect the presence of aircraft dispersal parking areas.

Detect a large cleared swath in a densely wooded area.

Detect, based on presence of piers and warehouses, a port facility.

Detect lines of transportation (either road or rail, but do not distinguish between).

Rating Level 2

Detect the presence of large (e.g., BLACKJACK, CAMBER, COCK, 707, 747) bombers or transports.

Identify large phased array radars (e.g., HENHOUSE, DOGHOUSE) by type.

Detect a military installation by building pattern and site configuration.

Detect road pattern, fence, and hardstand configuration at SSM launch sites (missile silos, launch control silos) within a known ICBM complex.

Detect large noncombatant ships (e.g., freighters or tankers) at a known port facility.

Identify athletic stadiums.

Rating Level 3

Detect medium-sized aircraft (e.g., FENCER, FLANKER, CURL, COKE, F-15).

Identify an ORBITA site on the basis of a 12-m dish antenna normally mounted on a circular building.

Detect vehicle revetments at a ground forces facility.

Detect vehicles/pieces of equipment at a SAM, SSM, or ABM fixed missile site.

Determine the location of the superstructure (e.g., fore, amidships, aft) on a medium-sized freighter.

Identify a medium-sized (approximately six-track) railroad classification yard.

Rating Level 4

Distinguish between large rotary-wing and medium fixed-wing aircraft (e.g., HALO versus CRUSTY transport).

Detect recent cable scars between facilities or command posts.

Detect individual vehicles in a row at a known motor pool.

Distinguish between open and closed sliding roof areas on a single bay garage at a mobile base station.

Identify square bow shape of ROPUCHA class (LST).

Detect all rail/road bridges.

Rating Level 5

Count all medium helicopters (e.g., HIND, HIP, HAZE, HOUND, PUMA, WASP).

Detect deployed TWIN EAR antenna.

Distinguish between river crossing equipment and medium/heavy armored vehicles by size and shape (e.g., MTU-20 versus T-62 MBT).

Detect missile support equipment at an SS-25 RTP (e.g., TEL, MSV).

Distinguish bow shape and length/width differences of SSNs.

Detect the break between railcars (count railcars).

Rating Level 6

Distinguish between variable and fixed-wing fighter aircraft (e.g., FENCER versus FLANKER).

Distinguish between the BAR LOCK and the SIDE NET antennas at a BAR LOCK/SIDE NET acquisition radar site.

Distinguish between small support vehicles (e.g., UAZ-69, UAZ-469) and tanks (e.g., T-72, T-80).

Identify SS-24 launch triplet at a known location.

Distinguish between the raised helicopter deck on a KRESTA II (CG) and the helicopter deck with main deck on a KRESTA I (CG).

Identify a vessel by class when singly deployed (e.g., YANKEE I, DELTA I, KRIVAK II FFG).

Detect cargo on a railroad flatcar or in a gondola.

Rating Level 7

Identify small fighter aircraft by type (e.g., FISHBED, FITTER, FLOGGER).

Distinguish between electronics van trailers (without tractor) and van trucks in garrison.

Distinguish by size and configuration between a turreted, tracked APC and a medium tank (e.g., BMP-1/2 versus T-64).

Detect a missile on the launcher in an SA-2 launch revetment.

Distinguish between bow-mounted missile system on KRIVAK i/II and bow-mounted gun turret on KRIVAK III.

Detect road/street lamps in an urban, residential, or military complex.

Rating Level 8

Distinguish the fuselage difference between a HIND and a HIP helicopter.

Distinguish between the FAN SONG E missile control radar and the FAN SONG F based on the number of parabolic dish antennas (three versus one).

Identify the SA-6 transloader when other SA-6 equipment is present.

Distinguish limber hole shape and configuration differences between DELTA I and YANKEE I (SSBNs).

Identify the dome/vent pattern on rail tank cars.

Rating Level 9

Detect major modifications to large aircraft (e.g., fairings, pods, winglets).

Identify the shape of antennas on EW/GCI/ACQ radars as parabolic, parabolic with clipped corners, or rectangular.

Identify, based on presence or absence of turret, size of gun tube, and chassis configuration, wheeled or tracked APCs by type (e.g., BTR-80, BMP-1/2, MT-LB, M113).

Identify the forward fins on an SA-3 missile.

Identify individual hatch covers of vertically launched SA-N-6 surface-to-air system.

Identify trucks as cab-over-engine or engine-in-front.

Table A.4
Civil NIIRS, March 1996

Rating Level 0

Interpretability of the imagery is precluded by obscuration, degradation, or very poor resolution.

Rating Level 1

Distinguish between major land use classes (e.g., urban, agricultural, forest, water, barren).

Detect a medium-sized port facility.

Distinguish between runways and taxiways at a large airfield.

Identify large area drainage patterns by type (e.g., dendritic, trellis, radial).

Rating Level 2

Identify large (i.e., greater than 160 acres) center-pivot irrigated fields during the growing season.

Detect large buildings (e.g., hospitals, factories).

Identify road patterns, like cloverleafs, on major highway systems.

Detect ice-breaker tracks.

Detect the wake from a large (e.g., greater than 300 ft) ship.

Rating Level 3

Detect large area (greater than 160 acres) contour plowing.

Detect individual houses in residential neighborhoods.

Detect trains or strings of standard rolling stock on railroad tracks (not individual cars).

Identify inland waterways navigable by barges.

Distinguish between natural forest stands and orchards.

Rating Level 4

Identify farm buildings as barns, silos, or residences.

Count unoccupied railroad tracks along right-of-way or in a railroad yard.

Detect basketball court, tennis court, volleyball court in urban areas.

Identify individual tracks, rail pairs, control towers, switching points in railyard.

Detect jeep rails through grassland.

Rating Level 5

Identify Christmas tree plantations.

Detect open bay doors of vehicle storage buildings.

Identify tents (larger than two person) at established recreational camping areas.

Distinguish between stands of coniferous and deciduous trees during leaf-off condition.

Detect large animals (e.g., elephants, rhinoceros, giraffes) in grasslands.

Rating Level 6

Detect narcotics intercropping based on texture.

Distinguish between row (e.g., corn, soybean) crops and small grain (e.g., wheat, oats) crops.

Identify automobiles as sedans or station wagons.

Identify individual telephone/electric poles in residential neighborhoods.

Detect foot trail through barren neighborhoods.

Rating Level 7

Identify individual mature cotton plants in a known cotton field.

Identify individual railroad ties.

Detect individual steps on a stairway.

Detect stumps and rocks in forest clearings and meadows.

Rating Level 8

Count individual baby pigs.

Identify a USGS benchmark set in a paved surface.

Identify grill detailing and/or the license plate on a passenger/truck type vehicle.

Identify individual pine seedlings.

Identify individual water lilies on a pond.

Identify windshield wipers on a vehicle.

Rating Level 9

Identify individual grain heads on small grain (e.g., wheat, oats, barley).

Identify individual barbs on a barbed wire fence.

Detect individual spikes in railroad ties.

Identify individual bunches of pine needles.

Identify an ear tag on large animals (e.g., deer, elk, moose).

Table A.5
Multispectral NIIRS, February 1995

Level 1

Distinguish between urban and rural areas.

Identify a large wetland (greater than 100 acres).

Detect meander flood plains (characterized by features such as channel scars, oxbow lakes, meander scrolls).

Delineate coastal shorelines.

Detect major highway and rail bridges over water (e.g., Golden Gate, Chesapeake Bay).

Delineate extent of snow or ice cover.

Level 2

Detect multilane highways.

Detect strip mining.

Determine water current direction as indicated by color differences (e.g., tributary entering large water feature, chlorophyll or sediment pattern).

Detect timber clear-cutting.

Delineate extent of cultivated land.

Identify riverine flood plains.

Level 3

Detect vegetation/soil moisture differences along a linear feature (suggesting the presence of a fence line).

Identify major street patterns in urban areas.

Identify golf courses.

Identify shoreline indications of predominant water currents.

Distinguish among residential, commercial, and industrial areas within an urban area.

Detect reservoir depletion.

Level 4

Detect recently constructed weapons positions (e.g., tank, artillery, self-propelled gun) based on the presence of revetments, berms, and ground scarring in vegetated areas.

Distinguish between two-lane improved and unimproved roads.

Detect indications of natural surface airstrip maintenance or improvements (e.g., runway extensions, grading, resurfacing, brush removal, vegetation cutting).

Detect landslide or rockslide large enough to obstruct a single-lane road.

Detect small boats (15–20 feet in length) in open water.

Identify small areas suitable for use as light fixed-wing (e.g., Cessna, Piper Cub, Beechcraft) landing strips.

Level 5

Detect automobile in a parking lot.

Identify beach terrain suitable for amphibious landing operation.

Detect ditch irrigation of beet fields.

Detect disruptive use of paints or coatings on buildings/structures at a ground forces installation.

Detect raw construction materials in ground forces deployment areas (e.g., timber, sand, gravel).

Level 6

Detect summer woodland camouflage netting large enough to cover a tank against a scattered tree background.

Detect foot trail through tall grass.

Detect navigational channel markers and mooring buoys in water.

Detect livestock in open but fenced areas.

Detect recently installed minefields in ground forces deployment areas based on a regular pattern of disturbed earth or vegetation.

Count individual dwelling in subsistence housing areas (e.g., squatter settlements, refugee camps).

Level 7

Distinguish between tanks and three-dimensional tank decoys.

Identify individual 55-gallon drums.

Detect small marine mammals (e.g., harbor seals) on sand/gravel beaches.

Detect underwater pier footings.

Detect foxholes by ring of spoil outlining hole.

Distinguish individual rows of truck crops.

APPENDIX B

Orbits and Geometrical Considerations

This appendix lays the foundation required to come up with prefiltering criteria for the targets, according to orbits and geometrical considerations. We start by briefly surveying the different representations of the earth and selecting a model that is appropriate for our goals of tasking ISR assets, in general, and generating collection decks, in particular. The appendix continues with an examination of the different systems of coordinates required for the analysis. We will see that the earth-centered, earth-fixed (ECEF) coordinate system is used to represent the locations of platforms and targets, that the east-north-up (ENU) coordinate system centered on the target is required to assess whether local obstructions hinder the view from the platform, and that a variant of the roll-pitch-yaw coordinate system affixed to the platform is indispensable for the computation of angles with respect to the platform. The important subject of calculating angles of grazing, depression, and squint is also addressed, as well as the procedure followed to determine whether free LOS is available between a target and an orbit point.

The first issue to address is the selection of one among the many models of the earth, of which the simplest is the "flat earth," which neglects the planet's curvature. The flat earth representation is adequate for situations in which the sensor and the target are relatively close in proximity. However, for the typical altitudes at which ISR platforms operate—between 30,000 and 60,000 feet—and for platform-to-target ranges that reach hundreds of kilometers, the curvature of the earth cannot be neglected, since it imposes a limit on the maximum range beyond which the LOS between the airborne platform and targets is blocked. Of course, this geometrical limit, referred to as the platform or sensor's horizon, does not occur when the earth is assumed to be flat. Figure B.1 shows an airborne sensor at a height H above the earth's surface (the platform's altitude) and a target located just at this sensor's horizon. Therefore, the LOS vector is tangent to the earth's surface at the location of the target. The LOS vector is defined as the vector with a starting point at the position of the platform that ends at the position of the target. Re is the radius of the earth (6,370 km). The angle between the LOS vector and the local vertical at the position of the target is 90 degrees. Therefore, the Pythagorean theorem yields

$$(Re + H)^2 = Re^2 + R_{max}{}^2.$$

After expanding the square and neglecting the H^2 term—permissible since H is much smaller than Re—we get the following expression for R_{max} in terms of H and Re:

$$R_{max} = \sqrt{2HRe}.$$

Figure B.1
Maximum Range, R_{max}, Due to the Earth's Curvature for a Platform Operating at Altitude *H*

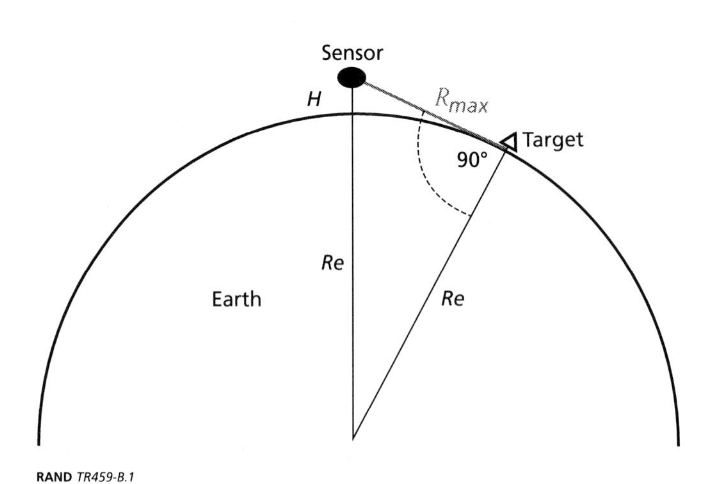

RAND *TR459-B.1*

This expression yields a maximum range of 341 km for a platform flying at 30,000 feet and of 483 km for an ISR asset located at 60,000 feet above the earth's surface.

The earth is not, however, a perfect sphere, and it is better approximated by an ellipsoid of revolution with its shorter radius at the poles and its longer radius at the equator. The most common models of the earth either globally or locally approximate the planet to one of several reference ellipsoids based on different survey data. The global ellipsoids deviate from a sphere by only plus or minus 10 km (to be compared with the earth's radius of 6,370 km), and locations on the earth referenced to different ellipsoidal approximations can differ from each other by 0.1 to 1 km (see Grewal, Weill, and Andrews, 2001, Chapter 6). An international standard called the World Geodesic System was set in 1984. This standard approximates the earth's surface at mean sea level by an ellipsoid of revolution with an axis that coincides with the earth's axis and a prime meridian that passes through Greenwich.

Insofar as modeling the ISR collection process is concerned, one of the most important aspects to consider for selecting a spherical or an ellipsoidal representation of the earth stems from possible differences in the grazing angle at the target's position. As we pointed out earlier, this angle has a significant effect on the quality of SAR images. The actual value for this angle is strongly dependent on the details of the local topography at the target's position—specifically, on the orientation of the target local horizontal plane, i.e., the plane perpendicular to the local vertical. The effect of the target local horizontal plane on the grazing angle is much more important than the smaller difference attributable to the exact shape assigned to the earth. Furthermore, in general, the orientation of such plane is not known. Therefore, to avoid unnecessary mathematical complications brought about by using an ellipsoidal representation, we decided to model the earth as a sphere. Another consequence of modeling the earth as a sphere is the error introduced in the absolute geolocation of targets and orbit points, which, as stated above, can vary from 0.1 to 1 km depending on the model selected. What is important for our analysis is finding not the absolute geocoordinates of targets and orbit points but the value of the platform-to-target range. We believe the error in range due to selecting a spherical representation for the earth to be significantly smaller than the error in absolute coordinates of 100 to 1,000 m.[1]

The second issue to address is the selection of the appropriate system or systems of coordinates to represent the locations of platforms and targets. One of the systems of coordinates most widely used to represent locations on or near the surface of the earth is the ECEF coordinate system, shown in Figure B.2. It has its origin at the earth's center of gravity and contains three mutually orthogonal axes, defined as follows:

- The first axis passes though the prime or Greenwich meridian; call it x-axis.
- The second axis generates a right-handed orthogonal system with the other two; call it y-axis.
- The third axis is aligned with the earth's polar axis; call that z-axis.[2]

ECEF represents the location of a point on or near the earth's surface by a triplet of coordinates—(x, y, z)—called Cartesian coordinates, which are also the components of the position

Figure B.2
ECEF Coordinate System

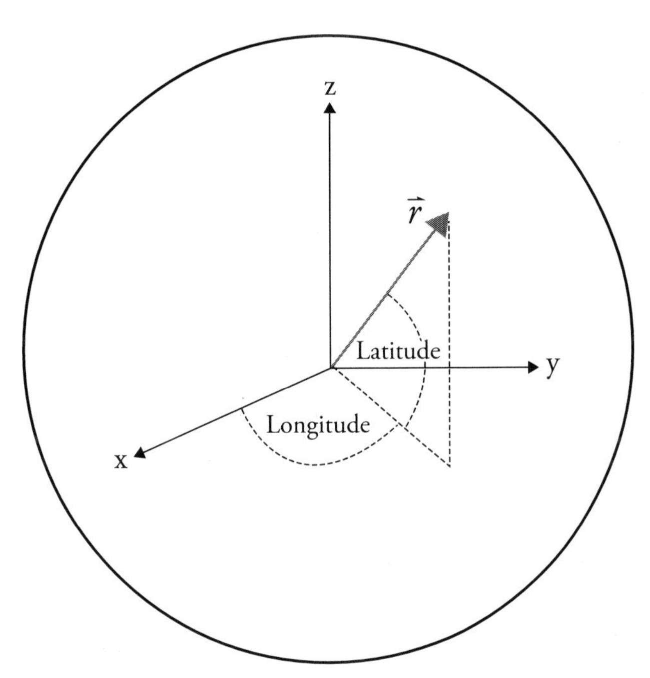

SOURCE: Grewal, Weill, and Andrews, 2001, Appendix C.
RAND *TR459-B.2*

[1] This statement is submitted without proof. It would be quite possible to quantify this error by distinguishing among geodetic, parametric, and geocentric latitudes and performing some tedious math (see Grewal, Weill, and Andrews, 2001, Figure C.6).

[2] To generate a right-handed orthogonal coordinate system, direct the index finger of your right hand toward the x-axis and the middle finger of the same hand toward the y-axis. Then, if the thumb is pointing upward, it should define the position of the z-axis.

vector for that point. Alternatively, the location of this point can be represented using polar or spherical (instead of Cartesian) coordinates by making use of a distance and two angles. The distance corresponds to the magnitude of the position vector, which is also equal to the earth's radius plus the radial extent above the earth's surface or altitude. The first angle, called latitude, is the complement[3] of the angle between the position vector and the polar or z-axis. The second angle, or longitude, corresponds to the angle between the x-axis and the projection of the position vector onto the x-y plane, which contains the equator.

These two sets of coordinates are related as follows:

$$x = |\vec{r}| \cos(Lat) \cos(Long),$$
$$y = |\vec{r}| \cos(Lat) \sin(Long),$$
$$z = |\vec{r}| \sin(Lat).$$

The magnitude of the position vector is related to the earth's radius and the altitude by

$$|\vec{r}| = Re + H.$$

In this paragraph, we briefly explain how we generated the orbits for the ISR platforms and how we calculated the grazing angle for each target–orbit point pair. Orbits for each one of the ISR platforms were drawn making use of information provided by an intelligence officer and Air Force fellow at RAND and using orbit polygonal envelopes taken from the Terminal Fury exercise. The detailed locations and shapes of these orbits are classified. From knowledge of the orbit and of the platform's speed, geolocations for points along the trajectory were generated at one-minute time increments using a simple program written in SEAS software. The steps that follow allowed us to determine the grazing angle of the LOS vector at the position of the target. First, the components of the target position vector and of the orbit point (or platform) position vector were computed from knowledge of their spherical coordinates—longitude, latitude, and altitude plus the earth's radius—using the mathematical formulae just given. The LOS vector was then calculated as the target position vector minus the platform position vector. Second, two angles were computed as illustrated in Figure B.3. These are the angle between the target and orbit point position vectors, labeled φ in the figure, and the angle that the LOS vector creates with the local vertical at the sensor or platform's local plane, called η, or nadir angle. Finally, the grazing angle was found from knowledge of φ and η via

$$\gamma = 90^\circ - (\varphi + \eta).$$

Locations of targets and ISR platforms on or near the surface of the earth are represented by means of Cartesian or spherical coordinates in an ECEF system, as previously explained. Modeling the collection process requires two additional coordinate systems to determine whether unobstructed LOS exists between the platform and target and to compute angles with respect to the platform's direction of flight.

---

[3] The complement of any angle β is defined as $90^\circ - \beta$.

Figure B.3
Determination of the Grazing Angle γ

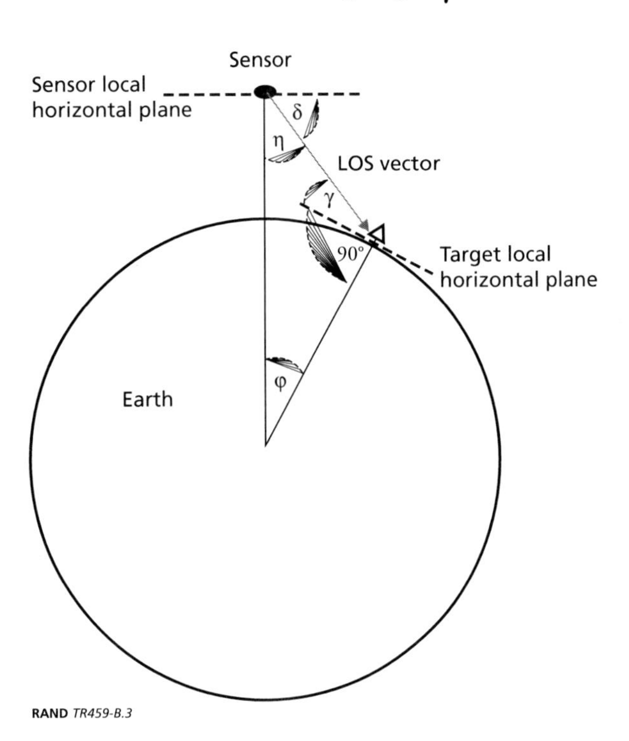

RAND *TR459-B.3*

A coordinate system affixed to the earth's surface at the location of the target is employed to assess whether local obstructions hinder the view from the platform. This ENU coordinate system (Grewal, Weill, and Andrews, 2001, Appendix C) has its origin at the target's location, and it has three mutually orthogonal axes that are aligned in the directions indicated by its name, i.e., east, north, and up (see Figure B.4). Just as for ECEF, a point is represented in ENU by its spherical coordinates, namely, by a distance (the magnitude of the position vector defined in the ENU coordinate system) plus an azimuth angle and an elevation angle. Azimuth is the angle between the east axis and the projection of the position vector onto the east-north plane, measured in the counterclockwise direction. Elevation is the angle subtended by the position vector and its projection onto the east-north plane. Notice that the east and north axes define a plane that is tangent to the surface of the earth, with the up axis being perpendicular to this plane. The plane defined by the east and north axes, in general, does not coincide with the local target plane, whose inclination is unknown. Hence, the up axis is not aligned with the true local vertical either. Nevertheless, insofar as LOS calculations are concerned, such misalignments are not relevant.

Let an arbitrary vector, V, be represented by ECEF coordinates V_x, V_y, and V_z. The components V_E and V_N of that vector in the target's local ENU coordinate system are related to its ECEF components by the following formulae:

$$V_E = -V_x \sin(Long) + V_y \cos(Long),$$

$$V_N = -V_x \cos(Long)\sin(Lat) - V_y \sin(Long)\sin(Lat) + V_z \cos(Lat).$$

Figure B.4
ENU Coordinate System, Centered at the Target's Location

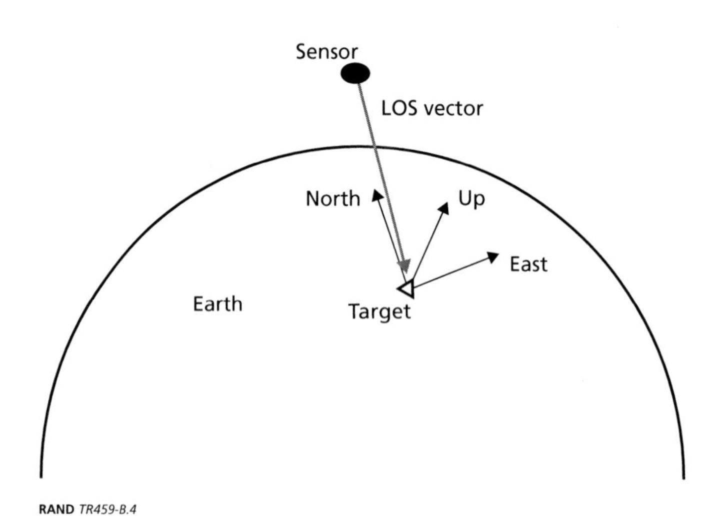

RAND *TR459-B.4*

V_E and V_N are also the components of the vector resulting from projecting vector V onto the east-north plane. Call the projected vector V_p. The angle subtended between such projection and the east axis is equal to the azimuth angle, and is given by:

$$\varepsilon_E = ArcCos(V_E / |\overline{V_p}|).$$

Notice that the denominator in the argument of the arc cosine is the magnitude of vector V_p.

The LOS calculations proceed in three different phases. The first phase creates a rectangular grid in the geographical area where the targets of interest are located and makes use of digital terrain elevation data. For each square of the grid, we define an ENU coordinate system. Paying attention to one of these grid squares only, points above the square can be represented in spherical coordinates by a distance, an azimuth, and an elevation. The result of this phase is the generation of masking angles as a function of azimuth for each square of the grid, where masking angle is defined as the minimum elevation angle above which LOS is unobstructed by the local topography. Azimuths are given in twelve 30-degree increments, from 0 to 330 degrees.

The second phase in the LOS calculations is the simplest of the three. It consists of assigning a grid square to each one of the targets of interest according to its geographical location.

In the third phase, we pay attention to the representation of the position vector of the platform (orbit point) in the ENU coordinate system affixed to the target.[4] By examining Figure B.4, we see that this vector is the same as the negative (or minus) of the LOS vector defined in the ECEF coordinate system. Moreover, the elevation angle of the platform in the target's ENU system is nothing more than the grazing angle calculated according to Figure

[4] To be precise, we need to distinguish two ENU coordinate systems; one has its origin at the target and the other one at the center of the corresponding square of the grid. If the rectangular grid is fine enough, these two ENU systems, for practical purposes, coincide.

B.3. The objective of this third and last phase is to compare this elevation (or grazing) angle with the masking angle for the specific azimuth that the projection of the platform position vector makes on the target's east-north plane. Only if the grazing angle exceeds or is equal to the corresponding masking angle can we ascertain that the view from the platform will be unobstructed. Therefore, what remains to be accomplished is the calculation of the azimuth. For each orbit point, a transformation of coordinates is effected on the platform-to-target LOS vector, so as to determine its components along the east and north axes—remember that this vector was originally defined within the ECEF coordinate system. The transformation of coordinates results in the formulae given above, which relate vector components in the ENU system to vector components in the ECEF system (Grewal, Weill, and Andrews, 2001, Appendix C). The azimuth angle of the platform with respect to the target is the angle subtended between the east axis and the projection of the LOS vector onto the east-north plane. As already stated, the final step consists of comparing the elevation (or grazing) angle with the masking angle for the given azimuth determined in the first phase. The target is given a "pass" label only if the former exceeds or is equal to the latter. These steps are repeated for each target–orbit point pair.

Bibliography

26 Air Intelligence Squadron (AIS), "ISR Assessment: The Way Ahead," briefing, Pacific Air Forces, undated.

AFOTTP 2-3.2. *See* U.S. Air Force, Air and Space Operations Center, 2004.

AFOTTP 3-3.6. *See* U.S. Air Force, Air Force Doctrine Center, forthcoming.

Air Force Instruction 13-1AOC. *See* U.S. Air Force, Deputy Chief of Staff for Air & Space Operations, 2005.

Bradley, Carl M., *Intelligence, Surveillance and Reconnaissance in Support of Operation Iraqi Freedom: Challenges for Rapid Maneuvers and Joint C4ISR Integration and Interoperability*, Newport, R.I.: Naval War College, February 9, 2004.

Brown, Jason, and Max Pearson, "Assessing ISR Effects," briefing, 32nd AIS, Air Operations Center, Intelligence, Surveillance, and Reconnaissance Division, Headquarters U.S. Air Forces in Europe, Ramstein, undated.

Chairman, Joint Chiefs of Staff, *Doctrine for Reconnaissance, Surveillance, and Target Acquisition Support for Joint Operations (RSTA)*, Washington, D.C.: Joint Chiefs of Staff, Joint Publication 3-55, April 14, 1993.

———, *Doctrine for Intelligence Support to Joint Operations*, Washington, D.C.: Joint Chiefs of Staff, Joint Publication 2-0, March 9, 2000.

———, *Command and Control for Joint Air Operations,* Washington, D.C.: Joint Chiefs of Staff, Joint Publication 3-30, June 5, 2003.

———, *Joint and National Intelligence Support to Military Operations*, Washington D.C.: Joint Chiefs of Staff, Vol. 1, Joint Publication 2-01, Doc. Call No.: M-U 40592, October 7, 2004.

Cordesman, Anthony H., *Intelligence Lessons of the Iraq War(s)*, Washington D.C.: Center for Strategic and International Studies, August 6, 2004.

Fitzpatrick, Teresa L., *Intelligence Campaign Planning: Deciding to Move Toward Effects-Based Intelligence Operations*, Fort McNair, Washington, D.C.: National Defense University, National War College, January 12, 2005.

Grewal, Mohinder S., Lawrence R. Weill, and Angus P. Andrews, *Global Positioning Systems, Inertial Navigation, and Integration*, New York: John Wiley and Sons, 2001.

Gross, A., C. Andrews, and S. Hovanessian, *Tactical Resolution Requirements Briefing—Briefing Prepared for Army Space Technology and Research Office (ASTRO)*, El Segundo, Calif.: The Aerospace Corporation, Army Space Systems Office, January 4, 1991.

ISR Assessment CONOPS Writing Meeting, organized by U.S. Air Force, Air Combat Command/IN, Nellis AFB, Nev., April 11–14, 2005.

Johnson, Daniel R., *An "ISR Strategy" for Joint Campaign Strategy, Planning, Execution, and Assessment*, Air War College, Maxwell AFB, Ala.: Air University Press, Maxwell Paper No. 34, September 2004.

Joint Doctrine, *DOD Dictionary of Military and Associated Terms*, Washington, D.C.: J-7, Joint Publication 1-02, April 12, 2001.

Joint Pub 2-01. *See* Chairman, Joint Chiefs of Staff, 2004.

Joint Pub 2-0. *See* Chairman, Joint Chiefs of Staff, 2000.

Joint Pub 3-30. *See* Chairman, Joint Chiefs of Staff, 2003.

Joint Pub 3-55. *See* Chairman, Joint Chiefs of Staff, 1993.

Lambeth, Benjamin S., *NATO's Air War for Kosovo: A Strategic and Operational Assessment*, Santa Monica, Calif.: RAND Corporation, MR-1365-AF, 2001. As of April 20, 2007:
http://www.rand.org/pubs/monograph_reports/MR1365/

Leachtenauer, Jon C., and Ronald G. Driggers, *Surveillance and Reconnaissance Imaging Systems,* Norwood, Mass.: Artech House, Inc., 2001.

Leachtenauer, Jon C., William A. Malila, John M. Irvine, Linda P. Colburn, and Nanette L. Salvaggio, "General Image-Quality Equation: GIQE," *Applied Optics*, Vol. 36, No. 32, November 10, 1997.

Lingel, Sherrill, et al., unpublished RAND research on tasking and employing ISR assets in a major theater of war scenario in SEAS.

Poss, James, "ISR Strategy to Task Process and Tool Requirements," briefing, USAFE, October 13, 2004.

Rhodes, Carl, Jeff Hagen, and Mark Westergren, *A Strategies-to-Tasks Framework for Planning and Executing Intelligence, Surveillance, and Reconnaissance (ISR) Operations*, Santa Monica, Calif.: RAND Corporation, TR-434-AF, 2007. As of August 13, 2007:
http://www.rand.org/pubs/technical_reports/TR434/

Rhodes, Carl, et al., unpublished RAND research on an initial examination of the USAF deployment for Operation Iraqi Freedom.

Shlapak, David A., *Shaping the Future Air Force*, Santa Monica, Calif.: RAND Corporation, TR-322-AF, August 2006. As of April 20, 2007:
http://www.rand.org/pubs/technical_reports/TR322/

Thaler, David E., *Strategies to Tasks: A Framework for Linking Means and Ends*, Santa Monica, Calif.: RAND Corporation, MR-300-AF, 1993. As of April 20, 2007:
http://www.rand.org/pubs/monograph_reports/MR300/

U.S. Air Force, Air and Space Operations Center, *Air Force Operational Tactics, Techniques, and Procedures 2-3.2*, Maxwell AFB, Ala.: Air and Space Operations Center, December 13, 2004.

U.S. Air Force, Air Combat Command, *Intelligence, Surveillance and Reconnaissance CONOPS*, Langley AFB, Va.: Air Combat Command, December 1, 2004.

———, *ISR Assessment CONOPS*, Langley AFB, Va.: Air Combat Command, draft, April 2005a.

———, *Intelligence, Surveillance and Reconnaissance (ISR) Assessment Functional Concept*, Langley AFB, Va.: Air Combat Command, draft, July 1, 2005b.

U.S. Air Force, Air Force Doctrine Center, *Intelligence, Surveillance, and Reconnaissance Operations*, Maxwell AFB, Ala.: Air Force Doctrine Center, Air Force Operational Tactics, Techniques, and Procedures (AFOTTP) 3-3.6, forthcoming.

U.S. Air Force, Air Force Space Command, *Space & C4ISR Capabilities CONOPS*, Peterson AFB, Colo.: Air Force Space Command, September 29, 2003.

U.S. Air Force, Deputy Chief of Staff for Air & Space Operations, *Operational Procedures—Aerospace Operations Center*, Washington, D.C.: Air & Space Operations, Air Force Instruction 13-1AOC, Vol. 3, August 1, 2005.

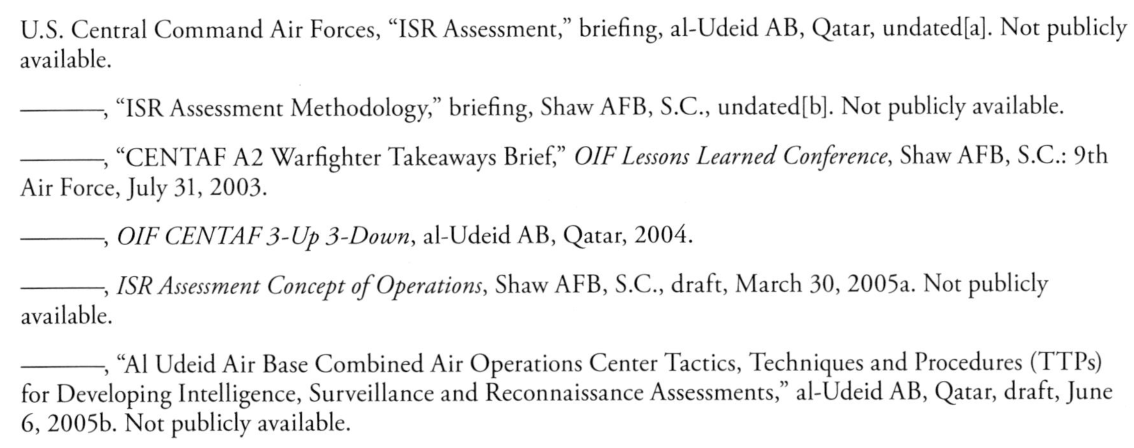

U.S. Central Command Air Forces, "ISR Assessment," briefing, al-Udeid AB, Qatar, undated[a]. Not publicly available.

———, "ISR Assessment Methodology," briefing, Shaw AFB, S.C., undated[b]. Not publicly available.

———, "CENTAF A2 Warfighter Takeaways Brief," *OIF Lessons Learned Conference*, Shaw AFB, S.C.: 9th Air Force, July 31, 2003.

———, *OIF CENTAF 3-Up 3-Down*, al-Udeid AB, Qatar, 2004.

———, *ISR Assessment Concept of Operations*, Shaw AFB, S.C., draft, March 30, 2005a. Not publicly available.

———, "Al Udeid Air Base Combined Air Operations Center Tactics, Techniques and Procedures (TTPs) for Developing Intelligence, Surveillance and Reconnaissance Assessments," al-Udeid AB, Qatar, draft, June 6, 2005b. Not publicly available.

U.S. House of Representatives, *Statement by Lieutenant General Duncan J. McNabb Before the Committee on Armed Services, Subcommittee on Terrorism, Unconventional Threats and Capabilities, Regarding Air Force Transformation,* Washington, D.C.: U.S. Government Printing Office, February 26, 2004. As of May 20, 2005:
http://www.globalsecurity.org/military/library/congress/2004_hr/040226-mcnabb.htm

U.S. Marine Corps, *Operation Iraqi Freedom (OIF): Lessons Learned*, June 2003.

U.S. Strategic Command, *Joint Functional Component Command-Intelligence, Surveillance and Reconnaissance (JFCC-ISR) Concept of Operations*, Omaha, Neb.: U.S. Strategic Command, draft, March 2, 2005.

Vick, Alan, Richard M. Moore, Bruce Pirnie, and John Stillion, *Aerospace Operations Against Elusive Ground Targets*, Santa Monica, Calif.: RAND Corporation, MR-1398-AF, 2001. As of April 20, 2007:
http://www.rand.org/pubs/monograph_reports/MR1398/

Zwicker, Steve, "ISR Assessment: HQ PACAF View," briefing, Pacific Air Forces, Hickam AFB, Hawaii, January 17, 2005.